Among the Ashes

Mennonite Reflections

Mennonite Reflections is dedicated to the publication of materials that explore the Mennonite experience. The series is published through the Institute of Anabaptist and Mennonite Studies, Conrad Grebel University College, University of Waterloo, in cooperation with Pandora Press.

The Silence Echoes: Memoirs of Trauma and Tears
Edited and translated by Sarah Dyck, 1997

Lifting the Veil: Mennonite Life in Russia Before the Revolution
Jacob H. Janzen. Edited with an introduction by Leonard Friesen.
Translated by Walter Klaassen, 1998

A Life Displaced: A Mennonite Woman's Flight from War-Torn Poland
Edna Schroeder Thiessen, 2000

Mennonite Alternative Service in Russia: The Story of Abram Dück and His Colleagues 1911-1917
John Dick and Lawrence Klippenstein, 2002

The Plain People: A Glimpse at Life among the Old Order Mennonites of Ontario
John F. Peters, 2003

From Kleefeld with Love
John A. Harder, 2003

Old Order Mennonites of Ontario: Gelassenheit, Discipleship, Brotherhood
Donald Martin, 2003

Pioneers in Ministry: Women Pastors in Ontario Mennonite Churches, 1973-2003
Mary A. Schiedel, 2003

A Family Torn Apart
Justina A. Neufeld, 2003

Between Worlds: Reflections of a Soviet-born Canadian Mennonite
Harry Loewen, 2006

Strange and Wonderful Paths: The Memoirs of Ralph Lebold
Ralph Lebold, 2006

Among the Ashes: In the Stalinkova Kolkhoz
Edited, translated and commentary by Peter J. Rahn, 2011

Among the Ashes

In the Stalinkova
Kolkhoz
(Kontiniusfeld)
1930-1935

Edited, translated, and commentary
by
Peter J. Rahn

Pandora Press

Library and Archives Canada Cataloguing in Publication

Among the ashes : in the Stalinkova Kolkhoz (Kontiniusfeld), 1930-1935 / edited, translated, and commentary by Peter J. Rahn.

(Mennonite reflections, ISSN 1480-3895)
Translated from the German.
The letters collected in this book form the correspondence received and preserved throughout Jacob Peter Rahn's lifetime.
Includes bibliographical references.
ISBN 978-1-926599-21-2

1. Rahn family--Correspondence. 2. Mennonites--Ukraine, Southern-- Biography. 3. Mennonites--Ukraine, Southern--History--20th century.
4. Farmers--Ukraine, Southern--History--20th century. 5. Collectivization of agriculture--Ukraine, Southern--History--20th century. 6. Ukraine, Southern--History--20th century. I. Rahn, Peter J. (Peter Jacob), 1939- II. Rahn, Jacob Peter, 1910-1991 III. Series: Mennonite reflections

DK508.9.S68A46 2011 947.7'10840922 C2011-905150-8

Book and cover design by Christian Snyder

PANDORA PRESS

AMONG THE ASHES.
ISBN 978-1-926599-21-2
ISSN 1480-3895

www.pandorapress.com

21 20 19 18 17 16 15 14 13 12 4 5 6 7 8 9 10

In memory of my parents

Jacob Peter Rahn (1910 – 1991)
and
Mary Adeline (Wiebe) Rahn (1911 – 1988)

If one of the great does something noble, all the writers mention it, but it seems to me that if one being small achieves fine deeds, to display this is even more worthwhile.

Xenophon, *Hellenica*

Table of Contents

1930 12 02 Constantly more is being demanded 136

– Hans is dealing in Rabbits, tends cows
– Henry is tending horses
– loft is empty, more is demanded
– neighbour [Aron Duerksen] is lazy but our children really love the sisters
– how will we celebrate Christmas
– wedding in Sparrau tomorrow
– J. Rahn's Anna with Heinrich Thiessen
– even the paper for letter writing is taken from us.

1930 12 15 How we will celebrate the Holidays we do not know 140

– winter is here, wind ripped tiles from the roof
– Papa had to go to Halbstatt last week
– even the straw is being collectivized
– Dirksen naeighbours are scattered, some arrested others just disappeared, children left behind

1930 12 22 May this separation unite us even more strongly 143

– the verse was a gift to me in these difficult days
– mail is not getting through, you and Abe should always write about both of you so the news gets here one way or the other

1930 12 28 Managed to spend 3 days of holiday in complete peace 145

– last duck in the oven
– family members have gone to church
– Susa[n] Toews took charge of the Christmas eve program through the help of the youth. She probably had nothing to lose, from one of the exiled families
– no dry goods of any kind.
– still thankful to sleep in our own beds
– thank you for the pictures
– glad you are learning well, but don't praise yourself too much.
– no paper, dollars being confiscated.
– famine will be far-reaching
– Henry wants to join MTS
– he and Hans have very poor clothes
– people being sentenced

Preface

In the Preface to her diaries, entitled *Hundreds and Thousands*, Emily Carr explains that she named it thus *because it* [her life recounted in her diary] *is made up of scraps of nothing which, put together, made the trimming and furnished the sweetness for what might otherwise have been a drab life sucked away without crunch.* She goes on to say that the title is actually the name of small English candy ... *that separately ... are not worth eating* ... but taken together provided a lovely and crunchy sweetness. Thus the experiences recorded in her diary, *too insignificant to have been considered individually, ... the little scraps of and nothingness of [her] life, have made a definite pattern.*[1]

While the collection of letters presented in this book might at first glance be seen more as the scraps of a bitter pill, out of the reading of these "scraps" emerges admiration for the determination of the writers of these letters to get on with life in some very devastating and debilitating circumstances and to find what little sweetness life might still afford them. The material collected here forms the correspondence received and preserved by one man, my father, Jacob Peter Rahn, [JP 1910] throughout his lifetime. His life spanned approximately 82 years (1910 – 1991), almost an entire century.

There are many reasons for undertaking this task, but two seem to stand out as I begin. First and foremost, the correspondence represents a documentary source for the historian. The Twentieth Century is a period of extraordinary change and momentous events. Moreover, my father did not have the luxury of growing up, working and dying sedately in a single community surrounded by family and friends. Like many others, he was born in a quiet agricultural village in the southern Ukraine (then called Kontiniusfeld, now Dolgoye). As a teenager, he endured the ravages of war, revolution and refugee dislocation. His adult life was spent in western Canada raising a family and trying to keep up with the rapid social, economic and technological changes that gradually turned the world into a global village. Thus the letters that he received and preserved represent an interesting and useful source of information for the cultural historian of the Twentieth

[1] Emily Carr, *Hundreds and Thousands* (Vancouver 1978), p. v. I thank my friend, Tarry Grieve for bringing this passage to my attention.

Century. This continuous correspondence written and mailed during a five-year period documents how many of the changes that ideological, political and economic forces brought to the Ukraine in the 1930s affected the ordinary human being, the *Everyman*.

Much of the written and photographic work telling the Mennonite story in the Ukraine and elsewhere in Southern Russia and later in the Soviet Union focuses on the fate of the political, religious and economic leadership of the Mennonite Communities. However, the letters written to my father emerge from a time when there is diminishing documentary material on what actually transpired in the Mennonite villages during the so-called collectivization and industrialization of agriculture as the five year plan set out in 1929 was being instituted throughout the Soviet Union. Terrible as the suffering of the Mennonite leadership was and stupid as the resultant destruction of the economic infrastructure was, the importance of the letters presented here lies in the fact that they convey the grass roots experience of the very ordinary people that communist ideology claimed to be liberating. The Rahn family continued to live in the same house from 1930 until 1942 and so experienced the entire spectrum of events in their village during the period of *dekulakization* and collectivization. These are not extraordinary individuals accomplishing extraordinary things but merely people desperately trying to survive and even seeking, usually under duress, to contribute to the new system through their work and energy. That many of them did not survive or experienced a fate beyond what any human being should be expected to endure, is a searing condemnation, if not of the entire communist ideology, at the very least, of the methods employed in the institution of that utopian idealism and of its accompanying propaganda.

Second, the letters may prove of value to JP 1910's children, grandchildren, siblings and their offspring, and to a wider audience, in contemplating the human identity and *raison d'etre*. The correspondence is centered on the life of one family and their experiences. As such it reflects a sense of extended family commitment and involvement that far surpasses anything that the modern nuclear family in its dislocation and isolation seems capable of achieving. Under what in many instances were severe difficulties, Jacob Rahn's family seemed driven to write, often even through anonymous unsigned postcards.

By today's standards they were not particularly well educated, but the letters speak eloquently of the sense of value that they placed on their relationship with each other. The members of this family, over time, found themselves living in vastly different circumstances from one another. Yet their respect, care, commitment and faith in and to each other endures. Their love was rooted in a deep and abiding belief in an almighty and loving God, a faith that seems to have been largely unshakeable, though it was severely tested.

When my father [JP1910] died in 1991, he left behind approximately two-hundred and fifty letters that he had received from 1930 until his death. Obviously these letters were highly important to him. When my aunt, Anni Rahn Wiens [1927], asked him if she could have some of the letters, he refused with the comment that these were an heirloom (*Erbgut*). Many of these were written long after he had emigrated to Canada and the rest of his family had left the Ukraine and been scattered throughout the Soviet Union or other parts of Europe. The translation and commentary focus mainly on the letters and postcards from 1930 to 1935. Though the later letters [1946 to 1991] reflect a dark and difficult period in Mennonite and Soviet history, there is considerably more documentary material of the sort in those letters that has been published from other sources. Thus the decision was made to restrict this volume to the earliest letters and postcards [fifty-four in all] that reflect a momentous and less documented period of both family and what might be considered Ukrainian Mennonite history.

To convey accurately in English what is contained in these letters is not always easy. The German was usually in the Gothic script. Thus the first step was to transliterate the cards and letters into Latin script. To do this I have relied on the help, first and foremost of my cousin, Christel [Prei] Stauffer of Leutenbach, Germany, and also my aunts, JP 1910's sisters, Anni [Rahn] Weins and Lena [Rahn] Pauls, and Mr. H. Woelk who, before he died, transliterated several of the postcards for a large gathering of Rahns and their offspring held in Germany in 1990. Wilf Hein of Abbotsford transliterated most of Postcard 1930 04 04 and Mary Epp also helped me with a few very difficult passages in the letters. Most of the rest I have transliterated myself. These letters were written in a somewhat colloquial peasant German that I also learned as a child at home. Thus I have usually tried to maintain the spelling and syntax found in the original. I have, however, standardized the dating of the letters to facilitate their proper chronological succession.

The next step of course was to translate these letters and cards into English. As a trained Classicist I have spent much of my life struggling with conveying in English the meaning of Greek and Latin texts for students who could not read the original and other academics who often could. In the German language the use of *und, aber, nun, dann, auch, wo, da, seit* and other similar connectors by the writers of these letters creates linked thought patterns that are often difficult to translate into smooth and literate English. In order to preserve the nuances of some of these German sentences I have frequently adopted a rather literal translation, even going so far as to try to maintain the original word order to create some sense of the thought pattern. Where that has resulted in excessively long, run-on sentences or jibberish in English, I have, with hesitation, broken the English sentences into smaller units and also resorted to extra comment in [square brackets]. While this method means that the English translations do not always read smoothly, it is hoped that a clearer understanding of the original nuance will emerge.

Within the body of these letters there is often reference to other members of the village and their fate under the surge of collectivization. Frequently, they are referred to by initials only. Since such information might well be useful to the relatives of other Mennonite families seeking further knowledge of the time and place of events like arrests, resettlement, exile to Northern or Eastern Russia or banishment from the villages of Kontiniusfeld and Sparrau, the footnotes or commentary suggest some possibilities as to who these might be. They are guesses only. However most of them are based upon four sources of information. First and foremost, my father paraphrased parts of many of these letters in his retirement years and often identifies in writing who is being referred to. Second, my two surviving aunts, Helena Rahn Pauls 1914 and Anni Rahn Wiens 1927 have been able to identify a few of them, though the former was not in Kontiniusfeld for most of the 1930s and the latter was only a toddler in 1930. Third, they and my deceased Aunt, Justa Rahn Prei Geddart 1922 have all published autobiographies that convey some information. Finally, there are village maps giving the residents of the various houses in both Kontiniusfeld and Sparrau from varying times in the 1920s and 30s published in the first edition of William Schroeder, *Mennonite Historical Atlas* (Winnipeg, 1990). These were supplemented by and compared with similar hand drawn maps published in my Aunt's autobiography, Anna Wiens, *Gepriesen sei die Gnade Gottes* (1991).

I would also be remiss not to mention the large volume of work done by my sister-in-law, Darlene Rahn, both in compiling family genealogy and in helping me with the formatting of this book. In addition, my brother, John Rahn, spent many hours scanning and formatting most of the documents, diagrams, charts and pictures found in this book. I am also grateful to the BC Mennonite Historical Society for the excellent facilities in which to consult the *Bote*, the *Rundschau* and other books for historical material to flesh out the context of these letters. Still in a sense this book is, in spite of commentary on the letters, incomplete. To my regret, I am unable to read Russian and so have not been able to make use of some of the archive material found in the Ukraine and in Moscow. However, both John B. Toews and Colin P. Neufeldt have been generous in providing me with photocopies of documents, translations, reference material and advice to fill this gap in part. I am also grateful to my brother-in-law, Jakob Sawatzky, who read and translated some of the Russian documents for me. Perhaps the letters published here will provide some further opportunity to these and other scholars to delve more deeply into the events of the 1930s in the Ukraine.

I am also most grateful to my cousin, Henry Henry Rahn, and his wife, Rita, who toured the former Mennonite villages of the Ukraine with me and my wife, Nancy, and provided valuable chauffeur and translation services. It gave me a much greater insight into what the Mennonites and my father's generation in particular had given up as a result of the revolution. Finally, I should also mention the much appreciated support and suggestions I received from Christian Snyder and Angie Hostetler at Pandora Press. To all the people mentioned above I express my deepest gratitude. However any errors that occur in this book are, as editor, translator and commentator, my responsibility. Last but not least, I thank my wife, Nancy, and my family for their interest, support and forebearance while I worked on this book.

The Historical Background

The Historical Background to the Letters, Dekulakization and Collectivization

In April of 1929 the sixteenth party conference of the Soviet Union approved a five-year plan that had, as one of its goals, the collectivization and industrialization of agriculture that was to include at least twenty percent of all peasant farmer households in the Soviet Union. In October 1929 Stalin and his closest advisors decided to change that goal to an all-out collectivization effort because according to his own written words (Pravda, Nov. 7, 1929) socialism had made a huge breakthrough and collectivization had been so enthusiastically supported by the peasant farmers. In January 1933 it was announced that the Five-Year Plan had, in fact been fulfilled after four years and three months. It was, however, in many ways a house of cards. During that four and one half year period food production [which was supposed to increase dramatically] actually went down, and in the process of trying to feed the Soviet population more than fifty percent of the farm animals (horses, cattle, sheep, goats and pigs) were slaughtered and millions of people lost their lives.[1]

The methods used to bring about this enforced collectivization in the Ukrainian farm villages are demonstrated in detail in an eyewitness account, *Execution by Hunger: The Hidden Holocaust*, by Miron Dolot (a nom-de-plume). In this book the author who lived through these years as boy in a Ukrainian village recounts the increasingly harsh steps used by party officials and the Red Army to force his village into a collective economy from 1930 to 1933. It is not an easy book to read. As a friend of mine who read it declared, reading this account is "really more than the soul can bear." Stalin's announcement in October 1929 meant in reality that Stalin was turning his back on the long held Communist expectation that peasant farmers and the proletariat would be a united force for establishing the ideal communist utopia. The peasant farmer particularly was now forced into the collectivization process, whether he wished to be part of it or not. The means employed to make this

[1] McClellan, pp. 120-130; Fitzpatrick 1994, pp. 39-41.

happen were probably similar, though not identical nor followed precisely the same order, throughout the Ukrainian villages whether the inhabitants were Ukrainians, Jews, German Lutherans or Catholics, or Mennonites.[2]

In 1930 the first step in subjecting the Ukrainian village to the collectivization process was to send in a gang of approximately ten men headed by a Twenty-five Thousander. This title designated him as a member of what was perhaps the most idealistic and zealous band of Communist party members. The others were party members or from the youth organizations of various cities throughout Soviet Russia. This group had unlimited power, carried weapons and were answerable only to the Central Committee of the Communist party of the Ukraine (headed for much of this time by Nikita Kruschev). They immediately arrested fifteen of the leading men in the village, evicted their families from their homes and deported them by train (probably to Siberia, although since none of them ever returned to the village, Dolot is not certain where they ended up). The houses of the evicted families became the offices and living quarters of communist officials. At the same time the village population was divided into groups of 100 people (Hundreds) and subdivided further into Tens and Fives. Each unit was headed by an individual functionary and agitator and all were subject to the Twenty-five Thousander and his propagandists. This created a large local bureaucracy of semi-literates (whose primary qualification was party membership) that replaced any past form of local self-government. To give this new structure the appearance of being a grass roots movement most policy decisions taken and implemented locally were usually ratified by a meeting of the local soviet, made up in theory of all the local villagers.

The farmland of the exiled peasants, who were of course nonparticipants in any of the meetings that ratified the actions against them, was officially designated a Kolkhoz, that is a collective, owned, in theory, by all the people of the state. In practice this meant that all meetings to decide local issues could be and often were attended by any party members from any where in

[2] It appears that in 1928 Stalin himself did not have an exact picture of what the kolkhoz entailed or should look like. See Fitzpatrick 1994, p. 39.

the Soviet Union, while locals who opposed the suggestions of the local party boss could often find themselves overridden in the decision making process and disenfranchised as enemies of the state. By 1933 measures such as these had created an extreme famine in the Southern Ukraine in which at least six million Ukrainians as well as many Mennonites died. Many of these Draconian measures were part of an ever-increasing deprivation of the private small farmer to overcome resistance to collectivization that had its origins in the nationalism, culture and mores of several centuries of Ukrainian village life.

The Mennonite Colonies of the Ukraine also had a long history of rather independent political and religious life. Beginning with the Chortitza (or Old Colony) and the Molotschna colonies and branching out from there throughout Southern Russia the Mennonites had developed agricultural settlements and traditions of self-government, schooling, religious observance and self-sufficiency that had by the 1930s been evolving since the initial immigration of Mennonites from the Vistula Delta in Prussia began in 1789. By approximately 1900 these colonies and the many large independent estates owned by Mennonites were so entrenched in the economic system of Imperial Russia [particularly in the grain trade where the Ukraine was often referred to as the *Bread Basket of Europe*], that several scholars have coined the term *Mennonite Commonwealth* to describe this Mennonite presence.[3] By 1917 when the Communist revolution overthrew the Czarist regime the Mennonite community was already highly stratified in terms of its agricultural economic activity.

[3] See, e.g., E. K. Francis, "The Mennonite Commonwealth in Russia, 1789-1914: A Sociological Interpretation," *MQR* 25, 173-182; David G. Rempel, "The Mennonite Commonwealth in Russia: A sketch of its Founding and Endurance, 1789-1919," *MQR* 47, 259-308; 48, 5-54; James Urry, "The Mennonite Commonwealth in Imperial Russia revisited," *MQR* 84, 229-247. This paradigm of Mennonite settlement in Russia has also been questioned by some scholars. See Urry, *ibid*.

In fact, it might well be argued that the Mennonites in Russia had a much more nuanced socio-economic stratification than the rather simplified Russian imperialist structure.[4] At the pinnacle stood the large estate owners [over 1000], who owned from as little as 100 to as much as 100,000 dessiatines of land[5] and the industrialists who manufactured much of the machinery and the building materials to be used in the farming communities.[6] In addition there were the grain traders who often owned and operated large mills [about seventy steam powered mills][7] and ran shipping companies in some of the larger cities like Nikopol, Melitopol or Odessa. In the villages themselves there were large landowners, known as *Groswirtschaften*, who controlled up to 300 dessiatines and generally lived in the center of the main street that ran the length of the village. In most villages a grist mill owner, a store owner, and a school teacher/preacher formed another stratum of the socio-economic ladder. Then there were the small landowners [often with forty dessiatines or less of land] and *Anwohner* who usually lived in rented housing and hired out their labor to the village farmers, and, finally, the village poor, such as the cowherds.

After the Bolosheviks seized power in 1917 most of the Mennonite estates along with those owned by people of other ethnic backgrounds were either abandoned or confiscated by marauding bands of peasants.[8] Most notable for the Mennonites was the activity of Nestor Makhno and his murderous band that initially attacked the Mennonite estates and then later several of the Mennonite villages during and after the civil war that

[4] For a description of the class structure under the Czars see Fitzpatrick 1993, pp. 209-212.

[5] Helmut T. Huebert, *Mennonite Estates in Imperial Russia* (Winnipeg 2005) vii-ix.

[6] Epp, *Mennonite Exodus* p. 17 suggests that in 1914 six percent of all farm machinery in Russia was produced in Mennonite factories, thirty-eight factories were producing brick and tile and eight industrial firms employed some 1,744 people and had gross income of about three million rubles.

[7] *Ibid.*

[8] Fitzpatrick 1993, p. 211. See also Priscilla Roosevelt, *Life on the Russian Country Estate* (New Haven, 1995) pp. 329-330.

followed.[9] To what extent this was Bolshevik policy or simply spontaneous peasant uprising is not easy to say.[10]

The Bolshevik ideologues that were largely Marxist intelligentsia had a problem. Marx had foreseen that a dictatorship of the proletariat would evolve more or less naturally out of the industrialization process as the bulk of the population became industrial workers and took power from the Bourgeoisie in a centuries old class conflict. Thus the majority of the population was expected to develop shared values that would find expression in a government controlled for and by the working classes. But after the 1917 Revolution and then the Civil War, the Russia under Communist control contained approximately 180 million people and an industrialized working class of approximately four million[11] and even that class had largely broken up as much of the industrial system was left unworkable by the revolutionary attack on the management stratum. To remedy this situation, the Bolsheviks introduced a New Economic Policy in 1921 that allowed limited free enterprise with the hope that the Bolsheviks could form an alliance with the peasant class [the bulk of the Russian Population] and under the dictatorship of the party [rather than the dictatorship of the working class or proletariat] gradually create a Marxist-Leninist utopia through the industrialization and then collectivization of agriculture. In order to accomplish this it was necessary to identify the various socio-economic classes in the situation of the 1920s.

[9] See e.g., Dyck, Staples & Toews 2004.

[10] The official Communist explanation was that the revolution against the Imperial system was a spontaneous uprising of the peasants against a corrupt and oppressive system, and that this revolutionary movement was only kept on track by the decisive actions of Lenin and his proletarian supporters. See, e.g., Gorky 1975 p. 311. However, one of the first acts of the Bolsheviks after seizing power was to issue a decree that ratified the seizure of estates, the redistribution of lands and the abolition of private property in land, et al. See Fitzpatrick 1994, p. 23.

[11] Fitzpatrick 1993, p. 214. It is interesting to note that one of the last Mennonite estates [1931] to be abandoned was owned by a Franz Rahn. See Huebert 2005, p. 158. Whether he was a relative of our family is uncertain. Since this estate was already situated in Siberia, west of Omsk, it may have remained under the communist radar, so to speak for longer than in other parts of the Soviet Union.

The census of 1926 was used as a tool to accomplish this differentiation. It broke the population down into two groups, those who worked for a salary [proletariat] and those designated as proprietors in both urban and rural settings.[12] This designation of these two population groups became the *raison d'etre* for much of the discrimination against the bulk of the population for many decades after the publication of the census since only industrial workers and agricultural laborers were designated as proletariat, while even most of the agricultural peasantry was included among the proprietary class. After 1926, at least in the Mennonite villages, the communists seem to have recognized three main groups – the poor, the middle peasants [my aunts spoke of the *Mittelreicher*] and the kulaks, who were seen to be exploiters of the agricultural population and so took the place of the bourgeoisie in Marxist theory.[13] Since these descriptions were very pejorative in terms of the Bolshevik analysis of the population, they stigmatized almost all of the participants of the former socio-economic system and potentially made the vast majority of the citizens of the former Russia *enemies of the State*.

The intent was to drive a wedge between the poorer peasant elements and the socio-economic leadership in the many peasant villages. On the whole this was not a very successful strategy for the Communists in the Mennonite villages,[14] though it did result in many of the leading families selling what they had [at whatever prices they could obtain] and emigrating from the Soviet Union in the 1920s and, from the Bolshevik point of view, rid the state of some of the hated Kulaks. At the same time it also justified the *de facto* seizure of the various estates, privately owned industrial and manufacturing complexes and other private property. But the limited free enterprise of the NEP at the beginning of this decade also allowed my grandfather, Peter Jacob Rahn 1883, the legal right to obtain a small farm in Kontiniusfeld and to develop his own *Wirtschaft* [agricultural enterprise].[15]

[12] Op. cit., p. 218.

[13] Fitzpatrick 1993, pp. 213-218.

[14] See, e.g., the letters of my grandfather, 1930 02 19 & 1930 03 12. This seems to have been true in many non-Mennonite agricultural villages as well, see Fitzpatrick 1994, pp. 49-53.

[15] See chapter entitled *A Brief Rahn Family History* in this book.

A second issue that confronted the Mennonites and probably most other non-ethnic Russians was the Russification drive. This had already begun in the 1860s under the Czars, but now came to the fore as the Bolsheviks attempted to unify the potential working class that was to include all the agricultural peasants. In the Mennonite colonies this was largely seen as an anti-German putsch that took the form of gradually replacing Mennonites with Russians in the Mennonite institutions such as schools, hospitals, orphanages and sanatoria. To counteract this Russian and, now, Bolshevik policy the Mennonites formed a Union (or *Verbannt*) of the Citizens of Dutch Lineage[16] under the leadership, among others, of B. B. Janz 1877, my father's uncle. For the Mennonites this organization worked partly as a buffer against the worst excesses of Bolshevik economic policy and to find a solution to the large number of Mennonite refugees who had over run the villages after the confiscation of the estates and the civil war.[17] When the B. B. Janz 1852 family, under duress, emigrated in 1926, it provided an example and then a haven for my father, Jacob Peter Rahn 1910, when he emigrated in 1930. Again, however, my grandfather Peter Jacob Rahn 1883 did not fare as badly as might be expected. The fact that he had served as a medic for the Russian army in the First World War provided a significant statement of his loyalty to the Russian state. Moreover, his status as an *anwohner* prior to the revolution meant that he would hardly be the first to come to the attention of the local Bolshevik officials in the late 1920s and early 1930s when most Mennonites came under personal scrutiny by the GPU and other Bolshevik organs.

[16] See Toews 1975, pp. 89-275. The name has often been viewed as somewhat disingenuous, see, e.g., Epp 1962 p. 30, 39. Toews 1975 p. 85, 86 thinks that the name was the direct result of Communist demands to abandon religious terminology [*Mennonite*] for something more cultural in nature [*Citizens of Dutch Lineage*].
[17] Op. cit., p. 295.

Finally the nature of governing power after the revolution of 1917 evolved dramatically in the next decade. Because of the excesses and anarchy of the revolutionary activity of the supposed proletariat in 1917 the very industrial structure that was to support this new society had also been left unmanned as management was removed and often killed and the working class abandoned the factories to foment the revolution.[18] To return a state of normalcy to the Soviet Union, the *Dictatorship of the Proletariat* soon became the dictatorship of the party through its all-party congresses,[19] to allow industry to begin to function again. In practice, of course such congresses were also too cumbersome for efficient daily decision-making. Hence, the dictatorship of the party was gradually replaced by a series of committees, of which the most important was the new Central Committee's Secretariat and Politburo.

The Tenth Party Congress[20] at the beginning of 1921 seems to have been a key in the evolution of the Communist Party's power structure. Through various political machinations this congress elected a Leninist slate to the Central Committee, defeated various motions put forward by Trotsky and the trade unions, and finally, in the interests of party unity, disbanded the various communist groups and forbade further factionalism. This meant that only individuals, not groups could criticize party policy. Moreover, the Central committee now also had the right to remove any of its members for factionalist activity. Through to his death in 1924 Lenin never used this tool, though he did remove opposition to his policies through posting those who were too vocal to assignments far away from Moscow and St. Petersburg so that they could not effectively take part in any leadership power struggles.

[18] See Fitzpatrick 1993, p. 212.

[19] See, e.g., Gorky 1975, p. 569.

[20] For a brief discussion of this Congress, see Fitzpatrick 1982, pp. 91-93.

Stalin, who had become General secretary of the Politburo after Lenin's death, not only used this tool to remove opposition, but, when he had defeated Leon Trotsky for the party leadership, also exile and then later arrest, detention and even killing in the name of party unity. Most important among the examples of this type of political control was the expulsion of Trotsky himself from the Central Committee and the party in 1927 and then his exile from the Soviet Union in February 1929 and his murder in Mexico in 1940.[21] Thus by early 1929 Stalin had absolute control of the Communist party apparatus and could embark on the first five year plan which in the name of industrialization and the collectivization of agriculture instituted the war against the peasants. In this confiscation of private property the tools of arrest, exile and murder were used readily against anyone who was deemed to stand in the way of Soviet "progress." The census of 1926 and the division of the population into workers [very few] and owners [very many] made most of the Soviet population potential *enemies of the state* and so justified the arbitrary exercise of communist power at the whim of the General Secretary and his henchman.

The removal of the last vestiges of real debate within the Party as well as in the Politburo through the removal of Trotsky from the Party and country, was in many ways the death knell for land ownership by the agricultural peasant population. In 1928, after some ten years of political propaganda and pressure from the Bolshevik authorities to move toward collectivization, only about two percent of the peasant landholdings had actually been deposited in the some 33,000 collective farms throughout the Soviet Union.[22] Stalin was already on record as ascribing this disappointing support for collectivization to the activity and, indeed, the very existence of kulak elements in the peasant villages.[23] The hoped for and expected support for the collectivization process that had not materialised among the agricultural peasants was now going to be forced upon the entire Soviet

[21] Fitzpatrick 1982, pp. 116-119.
[22] Fitzpatrick 1994, p. 40.
[23] Op. cit., p. 29.

Union. The first step in that process was, of course, the dekulakization of the peasant villages soon to be followed by the eradication of religion. How that process evolved has already been summarily described earlier in this chapter.

According to the ancient Greek historian and founder of empirical study of the past, Thucydides,[24] a historical account is to provide a record either of great expenditure, great suffering or grand stratagems [policy]. In providing a more detailed understanding of the day-to-day effects on the lives of the ordinary people of the Stalinist regime's dictatorial policies and activities, this publication of the letters from Kontiniusfeld from 1930 to 1935 presents a microcosm of policies and economic machination that resulted in massive suffering for millions upon millions of people. Certainly that result is not what Marx and Lenin had in mind when they first expounded their theory of a dictatorship of the proletariat. Whether the suffering of the masses actually mattered to Stalin and his henchmen in the 1930s is doubtful.

[24] *History of the Peloponnesian War*, I. 1-23.

Ancestoral Charts

The Crest of the Rahn Family from Tiegenhof

Jacob Peter Rahn 1910 – 1991 Paternal Ancestry.

Jacob J. Rahn
b. 1750
d. 13 Feb 1820
Platenhof, Gross, Prussia

?
b. 1740
d. 25 Mar 1790
Platenhof, Gross, Prussia

Abraham Rahn
b. 20 Oct 1782
Platenhof, Gross, Prussia
m. 31 May 1811
d. 1815
Platenhof, Gross, Prussia

Helena

Abraham Dyck
b. Prussia

Heinrich Abraham Dueck
b. about 1785
Prussia
m. 11 Nov 1806
Moesland, Prussia

Helena Isaak
b. 1786

Gerhard Dueck

Jacob Gerhard Dueck
b. about 1763

Johann Gerhard Rahn
b. 2 May 1813
Orloff, Gross, Prussia
m. 1832
Molotschna, South Russia
d. 1845
Sparrau, South Russia

Margaretha Dyck
b. 1 Mar 1815
Moesland, Prussia
d. 17 Apr 1896
Pastwa, South Russia

Jacob Rahn
b. 5 Dec 1840
Elisabeththal, South Russia
m. about 1877
d. Aug 1920

Peter Jacob Rahn
b. 21 Feb 1883
Elisabeththal, South Russia
d. 28 Apr 1942
Asiatic Russia

Martin Jacob Dueck
b. about 1820

Anna
b. about 1800

Helena Dueck
b. 6 May 1849
Sparrau, South Russia
d. Aug 1920

Jacob Peter Rahn 1910 – 1991 Maternal Ancestry.

Cornelius "Cornels" Jantz
b. 5 Dec 1739
m. 1774
d. 6 Aug 1803

Trincke Ratzlaff
b. 29 Jun 1749
Przechowka, Schwetz, Prussia

Tobias Cornelius Jantz
b. 4 Feb 1783
Jeziorka, Schwetz, Prussia
m. 20 Oct 1808
Karolswalde, Volhynia
d. 25 Jan 1848
Karolswalde, Volhynia

David Unruh

Susanna Unruh
b. 1 Oct 1790, Prussia
d. 22 Jun 1850 Karolswalde, Volhynia

Heinrich Block
b. about 1801
m. 22 Nov 1827
Neusass, Culm, Prussia

Andreas Janz
b. 5 Feb 1746
m. 1804
d. 20 Feb 1820

Maria Goerz

Maria Janz
b. 24 Dec 1807
Montau, Schwetz, Prussia

Johann Johann Penner
b. about 1788
Prussia

Johann Johann Penner
b. about 1813 Pastwa, Prussia
d. Before 1869
Molotschna, South Russia

Heinrich Abraham Dueck
b. about 1785
Prussia
m. 11 Nov 1806

Helena Isaak
b. 1786

Aganetha Dick
b. about 1812

Margaretha Dyck
b. 1 Mar 1815 Moesland, Prussia
d. 17 Apr 1896 Pastwa, South Russia

Abraham Rahn
b. 20 Oct 1782
m. 31 May 1811
d. 1815

Helena

Johann Gerhard Rahn
b. 2 May 1813 Orloff, Gross, Prussia
m. 1832 Molotschna, South Prussia
d. 1845 Sparrau, South Russia

Benjamin Jantz
b. 7 Jan 1821
Karolswalde, Volhynia
m. 21 Jul 1850
Karolswalde, Volhynia
d. 25 Oct 1889
Konteniusfeld, South Russia

Eva Block
b. 28 Oct 1828
Karolswalde, Volhynia
d. 7 Jul 1879
Konteniusfeld, South Russia

Benjamin B. Janz
b. 4 Apr 1852
Karolswalde, Volhynia
m. 2 Dec 1876
Konteniusfeld, South Russia
d. 8 Jan 1932
Coaldale, Alberta

Maria Janz
b. 11 Jul 1888
Konteniusfeld,
South Russia
d. 24 Dec 1920
Konteniusfeld,
South Russia

Johann Penner
b. 15 Aug 1833
Pastwa, South Russia
d. 2 Mar 1917
Hillsboro, Kansas

Sarah Rahn
b. 30 Jan 1836
Pastwa, South Russia
d. 16 Feb 1867
Pastwa, South Russia

Helena (Vogt) Penner
b. 14 Apr 1854
Pastwa, South Russia
d. 8 Apr 1941
Coaldale, Alberta

Peter Jacob Rahn 1883 Siblings

Jacob Rahn
b. 5 Dec 1840
d. Aug 1920
m. Tina Vogt
d. about 1877

- Heinrich Rahn (1872–1932)
  m. Liese Ewert (b. 1874)
  m. Anna Tjart
- Jacob Rahn (2 Apr 1874–1933)
  m. Maria Toews (26 Sep 1876–1942)
- Andreas Rahn (8 Jun 1876–24 Sep 1929)
  m. Anna Siebert (20 Nov 1877–28 Feb 1967)

m. about 1877 Helena Dueck
b. 6 May 1849
d. Aug 1920

- Johann J. Rahn (1879–14 Mar 1938)
  m. Truda Gerhard Gossen
  m. 1936 Maria Koop (26 Feb 1896–1945)
- Margaretha J. Rahn (1880–1932)
  m. Kornelius Friesen (1876–25 Oct 1932)
- Peter Jacob Rahn (21 Feb 1883–28 Apr 1942)
  m. 1 Jan. 1908 Maria Janz
  (11 Jul 1888–24 Dec 1920)
  m. 29 Sep 1921 Aganetha Schartner
  (17 Feb 1889–6 Jul 1968)
- Anna J. Rahn (14 May 1884–27 Jan 1976)
- Martin J. Rahn (28 Feb 1887–3 Mar 1967)
  m. Katharina Giesbrecht
  (3 Mar 1886–28 Apr 1959)
- David J. Rahn (26 Oct 1889–20 Oct 1955)
  m. Helena Thiessen (19 Aug 1892–5 Oct 1979)
- Helena J. Rahn (b. 1891)
  m. David Dueck

Maria Janz Rahn 1884 Siblings

Benjamin B. Janz
b. 4 Apr 1852
d. 8 Jan 1932
m. 2 Dec 1876 Helena (Vogt) Penner
b. 14 Apr 1854
d. 8 Apr 1941

- Benjamin B. Janz
  (25 Sep 1877–16 Oct 1964)
  m. 25 Sep 1905 Maria Rogalsky
  (14 Apr 1879–24 Aug 1886)
- Heinrich Benjamin Janz
  (13Jan 1879–24 Aug 1886)
- Margaretha Janz
  (6 Jul 1880–1 Feb 1973)
  m. 28 Sep 1903 Johann Aron Toews
  (23 Feb 1876–2 Sep 1953)
- Jakob Benjamin Janz
  (13 Jun 1884–30 Aug 1972)
  m. 24 Aug 1908 Aganetha Wiens
  (11 Jan 1887–2 Dec 1971)
- Maria Janz
  (11 Jul 1888–24 Dec 1920)
  m. 1 Jan 1908 Peter Jacob Rahn
  (21 Feb 1883–28 Apr 1942)
- Heinrich Benjamin Janz
  (28 Aug 1891–10 Jan 1893)
- Helena Janz (b. 31 Dec 1894)
  m. 1 May 1921 Abraham Aron Toews
  (11 May 1887–17 Feb 1962)

Aganetha Schartner Rahn Siblings

Tobias Schartner
b. 21 Oct 1838
d. 15 Apr 1921
m. first spouse unknown
- Friedrich Schartner (b. about 1873)
- Anna Schartner (b. about 1876)
- Peter Schartner (b. about 1877)

m. 14 Sep 1857 Maria Ratzlaff
d. 7 Feb 1915
- Maria Schartner (b. 2 Aug 1887)
- Aganetha Schartner (17 Feb 1889–6 Jul 1968)
  m. 29 Sep 1921 Peter Jacob Rahn
  (21 Feb 1883–28 Apr 1942)
- David Schnartner (b. 3 Jul 1891)
- Abram Schartner (May 1894–9 Aug 1974)
- Isaak Schartner (7 Mar 1896–10 Jul 1922)

Jacob Peter Rahn 1910 Siblings

Peter Jacob Rahn
b. 21 Feb 1883
d. 28 Apr 1942
m. 1 Jan 1908 Maria Janz
b. 11 Jul 1888
d. 24 Dec 1920

- Jacob Peter Rahn (3 Jan 1910–Aug 1991)
  m. 24 Oct 1937 Mary Adeline Wiebe
  (21 Jul 1911–25 Dec 1988)
- Heinrich Peter Rahn (3 Jun 1911–20 Apr 19
  m. Apr 1948 Susanna Klassen
  (23 Mar 1911–14 Sep 1998)
- Johann Peter Rahn (15 Mar 1913–1944)
- Helena Peter Rahn (b. 5 Aug 1914)
  m. David Aaron Pauls
  (18 Nov 1911–3 1998)
- Maria Peter Rahn (27 Oct 1918–27 Jan 192

m. 29 Sep 1921 Aganetha Schartner
b. 17 Feb 1889
d. 6 Jul 1968

- Justa Peter Rahn (10 Aug 1922–1 Jul 2004)
  m. Herman Prei)b. 16 Sep 1895)
  m. David Geddert
- Mika Peter Rahn (b. 12 Apr 1924)
  m. 1947 Heinz Plett (b. 9 Nov 1923)
- Anna Peter Rahn (b. 11 Aug 1927)
  m. 1 Feb 1948 Rudolph Wiens
  (6 Mar 1924–30 Jan 1995)

Heinrich Wölk's Rahn Ancestry

Jakob Rahn
d. Jul 1920
m. Helene (geb Dyck)
d. Oct 1920

- Heinrich (1872–1932)
- Jakob (1874–1933)
  m. Maria Toews
- Andreas (1876–1929)
- Margareta (m. Friesen, d. 1932/3)
- Johann
- Peter (1883–1942)
- Anna (b. 1884)
- Martin (d. 1967)
- David (1889–1935)
- Helene (m. Dyck, d. 196-)

- Katharina (1899-1977)
  m. Franz Friesen (1904–1938)
- Johann (1901–1938)
  m. Mika Neufeld (b. 1900)
- Jakob (1903–1937)
  m. Justina Friesen (b. 1910)
- Heinrich (1905–1984)
  m. Justina Friesen (b. 1910)
- Helene (1907–1994)
  m. Heinrich Wölk (b. 1906) — Heinrich Wölk b. 1930
  m. Anna Strjapuchina b. 1929
- Isaak (1909–1937)
  m. Kathar. Penner (b. 1913)
- Maria (1911–1933)
- Kornelius (1913–19--)
  m. Sara
- Peter (1916–1941)
- Gertruda (1918–1993)
  m. Bernh. Görzen (1915–1988)

A Brief Rahn Family History

A Brief Rahn Family History

The Rahn family name appears in the annals of European history in Switzerland, France and Germany in the fourteenth century. The first mention [1381] of the family in Switzerland is connected with present day Freienstein. At that time they were serfs of the Lords of Randenburg. The name had various early spellings, Ron, Rön, Rom, Röm, Rohn, Ran, Rän, Rahan, Rahne and then, Rahn. The three most frequently used first names are Hans [or Johann], Heinrich and Rudolph. Occasionally a Jakob appears, but no Peter is found. In 1429 a Hans Ram purchased his citizenship in the city of Zurich and thus founded a family of free citizens. Shortly thereafter some achieved knighthoods, and the family obtained a coat of arms.[1] Within a century the family name achieves prominence in political and civic affairs in Zurich. In the late 1500s and early 1600s at least three *burgermeisters* [mayors] of Zurich were Rahns and the surname appears rather frequently in other civic and commercial contexts. In the 1600s there was a prominent Swiss mathematician and physicist Hans Heinrich Rahn [1622-1676] who seems to have been the inventor of [or at least one of the first to use] the symbol for division.[2]

From here the name seems to have spread to France. There are Rahnes associated with the Huegenots in the Champagne region of France in the 1700s who seem to have fled to England or America as a result of persecution. They dropped the e from the spelling of their surname and now are indistinguishable from the Rahns whose background is traced back to other parts of Europe. Since the Huegenots were French Anabaptists, it is likely that they also originated from or had some connection with the Anabaptists in Zurich. However, it should also be noted that in January 1613 in a disputation aimed at drawing the Swiss Anabaptists back into the Calvinist church, one of those arguing the Calvinist position was the Burgermeister, Hans Rudolph Rahn [1560-1627], who held that position for twenty years until his death.[3] He was very anti-anabaptist. In fact, he probably presided over the execution of the last Anabaptist martyr in Zurich

[1] Schnyder-Spross, Plate 12 following p. 446.

[2] Schnyder-Spross, p. 7, 114, 262, 445-446.

[3] Schnyder-Spross, p. 61-62.

[in 1614], Hans Landis. Thus it seems unlikely that our Mennonite origins are from Zurich.

There have been Rahns in the United States since the 1700s. Some of these clearly descended from the Swiss branch of the family.[4] Most North American Rahns, however, seem to have their roots in Northern Germany, particularly Hamburg. A search of northern Europe reveals a village called *Rahnsdorf* [Rahns' village], now a suburb of the borough of Treptow-Köpenick, a South East section of Berlin. Mention of this little fishing village on the shore of the Müggelsee, the largest lake within the city limits of modern Berlin, is already made in 1375 in the land book of the Emperor Charles IV. This area also contains remains of a Slavic castle from about 800 AD, an indication that this area was connected with Eastern Europe. At least one major battle was fought in this territory during the religious wars. In the 16th century this region was part of a Prussian state under King Friedrich I. Interestingly he commissioned what may have been the first Protestant chapel in the area as part of the Baroque palace that was built over the ruins of the earlier castle.

Anabaptists seem to have found support and protection in the cities and towns of the Hanseatic League that encompassed in its territory much of the North East region of what is today modern Germany [formerly called Pomerania] and would have included Rahnsdorf. Not surprisingly then, the Rahn family name is found in Anabaptist records, since Pomerania was a center of Anabaptist activity. Another possible origin for the Mennonite Rahns may be a family known to have lived in East Prussia in a village called Napratten. The originators of this family, a Bartholomaey Rahn and wife Anna, had their first child in 1626 and the last, a Petrus Rahn, on 27 January 1641.[5] Tempting as it might be to connect this notice with our first known ancestor [see next paragraph], these Napratten Rahns appear to be Catholic for Petrus' birth and christening seem to have followed very quickly one upon the other. The most likely origin of our branch of the Rahns appears to be Pomerania.

[4] Schnyder-Spross 1951, p. 505.

[5] www.rahnfamily.org/pedigree/names1.html

The first Rahn[6] to have connections with the Mennonites that I have found in the church records is a Peter Rahn[7] [hence forth designated the family Patriarch], baptised in the Tiegenhagen Mennonite church in West Prussia in about 1640.[8] In 1664 he appears again as a landowner in Tiegenhof.[9] In 1669 he seems to have purchased[10] substantial property in the same region that may have become the centre of the subsequent Rahn family growth and expansion. During the 1700s the Rahn name appears frequently in records of churches and land registries of villages in and around Tiegenhof such as Altendorf, Orlofferfelde, Petershagen, Platenhof and Tiegenhagen. In particular, Johann [1775-1855] and Peter Rahn[11] [1782-1876] served the Tiegenhagen Mennonite church as Deacons (diakon). These were common first names in our branch of the family. Given the practice of people in this time period to name their eldest son after their grandfathers and subsequent sons after uncles, it is likely that they descended from the same ancestral tradition as our family, though they are not direct ancestors. A separate line of Rahns also associated with the Mennonite Churches in the same area

[6] U. Dueck, *Die Familie Rahn von Tiegenhof*, p. ii, was of the opinion that Peter Rahn probably came from Pomerania.

[7] Through out this chapter I will use bold type to indicate my direct ancestors. Given the large number of names, often similar or the same, bold print may make it easier to keep track of our lineage.

[8] The *International Genealogical Index* no. 7233221 lists Petrus Rahn as having been christened 27 January 1641 in Napratten Ost Preussen. This indicates that the Rahns were already in Prussia at this time. B. H. Unruh, *Die niederländisch-niederdeutschen Hintergrunde der mennonitschen Ostwanderungen...*, p. ??? mentions Peter Rahn from Tiegenhof about 1650.

[9] Horst Penner, *Ansiedlung mennonitischer Niederlaender im Weichselmuendsgebiet von der Mitte des 16. Jahrhunderts bis zum Beginn der preussischen Zeit* (Weierhof, 1963), p. 47 note 2.

[10] These were actually long term leases, since no German or Dutch-speaking foreigners were allowed to own land in Prussia at this time.

[11] Adalbert Goertz, *List of Mennonite Ministers in Prussia*, (published online at Prussian Mennonite Genealogical Resources by MMHS). These were probably brothers who were both born in Altendorf and later lived in Tiegenhagen and Petershagen resp. where they died.

of the Vistula Delta frequently used the first name Isbrand or Isbrandt.[12] Whether these originate from a younger son of the original Peter Rahn, or not, is not certain. However, since they lived in the same region of Prussia as our branch of the Rahns, it seems likely that they are somehow related.

It seems that the Rahns both in Pomerania and in Prussia were less involved in agriculture and more in some form of industry. According to Anna Andres, a well know genealogist, the Rahn family had a number of large businesses throughout Germany and Pomerania in the 1600s but the family seemed very susceptible to Tuberculosis and much of it died away.[13] Although Patriarch Peter became a substantial landowner, subsequent Rahn families in official civic lists owned small properties of between one and approximately forty acres. Peter is listed as a cable maker (*Seiler*) in the records, i.e., a small industrialist. Since the primary purpose of the Mennonites settling in the Vistula delta was to drain the land and make it arable, the use of ropes and cables in the building of canals and locks would have created a ready market for Peter's wares and contributed significantly to his increasing wealth. Also significant is the connection to the Tiegenhagen Mennonite church that was largely comprised of Netherland Mennonites who had established Tiegenhagen as a market centre for the agricultural wares produced in the region[14] and so tended to be more entrepreneurial.

The earliest ancestor to whom we can establish a direct connection is Jacob Rahn, born in 1750 and deceased 13 February 1810 in Platenhof bei Tiegenhof. Some sources attach a middle initial "J" to him, which might make his father a Jacob or Johann born probably between 1710-1730. We know of a Jakob Rahn, born in 1669, who according to Ulrich Dueck might have been the grandfather of Jacob Rahn 1750.[15] Because of the time lapse between the two [some eighty years] and the practice of naming the eldest son after the grandfather, very prevalent in the Rahn family, his supposition seems reasonable. If so, that would provide a direct link to patriarch Peter,

[12] Ulrich Dueck, p. 2.

[13] Ulrich Dueck, p. i.

[14] Penner 1963, pp. 46-47.

[15] *Die Familie Rahn von Tiegenhof*, p. 297.

who was probably the father of Jacob 1669.

Jacob 1750 married three times. His first wife [name unknown] died 25 March 1790. He married his second wife, Katharina Klassen, on 10 October 1790. She died shortly after. On 11 December 1791 he married Maria Wiebe, a woman some twelve years younger than he. She bore him several sons. Abraham Jacob was born 20 October 1792 in Platenhof and died in 1815. Another son, Gerhard Jacob, was born supposedly in 1790 but, more probably, 28 October 1795.

It is here that the story becomes intriguing. As a result of the Prussian government's restriction on Mennonites buying land from Prussians in 1789, many, especially younger sons, followed the original emigrations of Mennonites to the Chortitza 1787/89 and Molotschna 1803/04 in subsequent years. Gerhard was one of Jacob's [1750] younger sons who would have been affected by the new regulation. On 8 June 1827 the Russian General consulate issued an immigration visa to Gerhard Rahn from Ohrlofferfeld for himself, his wife Sarah, a son Johann, thirteen years of age, a daughter Maria, two years old, and a brother Peter, aged twenty-six. In the list of Mennonite households founded in Russia in 1828, Gerhard Rahn's family of three males and four females is noted to have settled in Sparrau, Molotshna region.[16]

That Gerhard has moved first from Platenhof bei Tiegenhagen, the home of his father, to Orlofferfeld reinforces the notion that as a younger son there was no more land available to him from his father's estate. Since his wife Sarah was baptised in the Orlofferfeld church it is probable that Gerhard found some sort of opportunity either for land or housing with his in-laws, a not unusual way of dealing with financial and family matters at this time. More interesting is the inclusion of Johann aged thirteen as his son. He was born 2 May 1814 when Gerhard would have been about nineteen years of age. While it is not impossible for him to have had a son by that age, Gerhard was only baptised on 4 June 1815, [usually a prelude

[16] Peter Rempel, *Mennonite Migrations to Russia 1788-1828* (Manitoba, 2000), p. 201, 207.

to marriage among the Mennonites, at that time] and married Sarah Wall on 4 November 1821. Given the attitudes to marriage and sexual behaviour among the Mennonites at that time, a more likely scenario is that Johann was a nephew who was adopted by Gerhard. According to Ulrich Dueck[17] Johann was actually the son of brother Abraham who died in 1815 when Johann was still an infant. There is a suggestion by my aunts, Anni Wiens and Lena Pauls, though never confirmed by him that our great grandfather Rahn was an orphan. As often happens, family oral traditions often compress facts of the past. It is more likely to have been our great-great grandfather, Johann, to whom this applied, since our subsequent ancestry is very clear, well known and carefully researched by our father, Jacob Peter Rahn [1910].[18]

Before moving to a discussion of our 19th century ancestry, it is necessary to examine the birth date of Gerhard. The Mennonite Genealogical data base [Grandma or GM for short] in earlier versions listed two Gerhard Jacob Rahns, both born of Jacob Rahn 1750 and Maria Wiebe 1762. The first is supposed to have been born in 1790, the second in 1795. Both are supposed to have married a Sarah Wall, and both have been thought to move to Sparrau in the Molotschna. They are probably one and the same person. The disparity in birthdates probably results from the fact that in earlier versions of Grandma an entry had been made by someone who knew partially of the descendants of Jacob 1750 and Maria Wiebe. Later versions of the database included the research of Ulrich Dueck who, by virtue of his mother, grandmother or aunt being a Rahn[19] had garnered more precise knowledge of our ancestry. Finally, a Gerhard born in 1790 of Maria Wiebe makes no sense because she did not marry until December 1791. Moreover, in 1790 Jacob Rahn 1750 lost both his first wife [name unknown] 25th March, and his second wife, Katharina Klassen, shortly after October 10th. Unless the first Gerhard was born of the first wife, before she died naturally or during childbirth, and there is no record of this, 1790 is an impossible

[17] *Die Familie Rahn von Tiegenhof*, p. 301.

[18] He recounts his ancestry very precisely in his memoirs. He had several aunts and uncles in Coaldale of great age whom he consulted about the facts that he recorded.

[19] She was PJ1883's youngest sister, Helena 1891. She married David Dueck who lived in Sparrau in the 1930s.

birth date. Thus it seems most plausible that there is but one Gerhard Rahn born 1795 who is the child of Jacob 1750 and Maria Wiebe.

After Gerhard had settled in the newly founded village of Sparrau in the Molotschna colony in the Ukraine in 1828, his adopted son, Johann, my great great grandfather, married a Margaretha Dyck in 1832. He seems to have lived for sometime in Elisabethtal since some of his six children were born there. Only three survived infanthood, Sarah [born 30 January 1836], Helena [born 13 April 1839] and Jacob [born 5 December 1840]. Gerhard died in Sparrau in 1845. Since Gerhard had no sons when he died, it is likely that Johann inherited whatever property his adoptive father, Gerhard, had accumulated in Sparrau.

Jacob Johann Rahn 1840, my great grandfather, also spent some time in Elizabethtal. He married twice, first to a Maria Vogt who died about 1877 after giving birth to three sons, Heinrich 1872, Jacob 1875, and Andreas 1877. He also must have lived in Pastwa, a village not far from Sparrau, for his three sons are listed in the school register for 1883-84 as students there. After Maria died, he married Helena Dueck [born 1849, died August 1920] who bore him seven children: David 1878, Johann 1879, Margaretha 1880, Peter Jacob 1883, Anna 1884, Martin 1888 and Helena 1891 – obviously a very productive marriage. Jacob 1840 moved back to Sparrau in 1891 and died there in August 1920 within a month of Helena's passing, apparently of a Hernia problem.[20] Although his was not one of the *Gross Wirschaften* when he died, he did have at least two houses in Sparrau, one of which ended up in the hands of the sons of the first wife. The other house actually had a second small hut on it that was only torn down about 2002, although the main house had been destroyed in the early 1940s. Some of the children of the second marriage, Johann 1879 and his family, Anna 1884, and Helena 1891 and husband David Dueck and family inhabited this property in the 1920s and 1930s.

Before proceeding to a discussion of the next generation of our ancestors, Jacob Johann1840's sister, Sarah 1836, is of interest because she is actually the mother of my father's grandmother, Helena Vogt Penner Janz 1854

[20] According to my Aunt Lena, he died of what she called *Bruch*. I suspect this was a duodenal hernia.

and so is his great grandmother. Sarah 1836 gave birth out of wedlock to daughter Helena on 14 April 1854. By all reports the father was an Andreas Vogt whom, according to my Aunts, Sarah did not marry for he already had a wife and family. However, after marrying Johann Penner[21] she died 16 February 1867 at age thirty-one in Pastwa when Helena was almost thirteen years of age. Johann Penner remarried in 1869. Since Helena went by the surname Penner until on 2 December 1876 she married Benjamin B. Janz 1852, it is likely that she remained in the Penner family after her mother's death, now with two step-parents. Among Helena and Benjamin Janz' seven children was Maria [born 11 July 1888, died 24 December 1920 of Typhoid fever], our grandmother.

By all accounts my great grandfather, Benjamin Janz 1852, was very well liked by family and neighbours in Kontiniusfeld where he settled and owned and operated a windmill until he and his family emigrated to Coaldale, Alberta in 1926. He grew and cultivated a very fine fruit orchard in his back yard near the Kuruschan river. His children and grandchildren universally declared him to be a very kind and gentle individual, much beloved by all. Apparently, whenever he came to visit any of his children's families, he always had some fresh fruit or candies in his pocket. It became a game for the grand children to extract these from him. The habit must have been well entrenched, since his grandson, my father Jacob Peter Rahn 1910, seemed to copy this form of enticement with his many grandchildren as well.

The memories of Helena Vogt Penner Janz 1854 are somewhat different. She was a much sterner, curt and more demanding individual. In Kontiniusfeld she insisted that if the grandchildren came to see them, they must always come via the back door. She also is remembered as one who refused to give the grandchildren even a crust of bread during the famine of 1920/21.[22] On the other hand, during that year in which my grandfather's family was without a mother, she took in and cared for Aunt Lena 1914 and Uncle Hans 1913 until my Grandfather, PJ 1883, remarried. After both the Janzes and my father had emigrated to Southern Alberta, the letters make clear that

[21] Johann Penner, b. 15 Aug 1833, died in Hillsboro Kansas 2 March 1917. According to some reports he married Sarah in 1855.

[22] It may well be that she had nothing to give.

my father was heartily welcomed into their home. She is always referred to with respect and affection. It is also evident that she wrote several letters to the P. J. Rahn family in Kontiniusfeld, even though her own daughter, the first mother of the family, had died some ten years earlier. Clearly, she never abandoned them but continued to encourage and support them emotionally, and together with her husband sent them money.

Peter Jacob Rahn 1883, my grandfather, who was born in Alexandertal, also spent part of his childhood there. He married twice. His first marriage to Maria Janz 1888 on 1 January 1908 produced five children – my father, Jacob [3 January 1910], Henry [3 June 1911], Johann [15 March 1913], Helena [18 August 1914] and Maria [27 October 1918, died 27 January 1921]. Though my father, the oldest in the family, was born in Kontiniusfeld, and at the beginning of the marriage his parents probably lived with my grand parents, they soon moved to Alexandertal to live on a farm rented from Rev. John A. Toews. During World War I PJ 1883 was conscripted into the medical corps beginning late in 1914.[23] He was a prisoner of war for almost three years in Germany. When he returned home after the war he was immediately thrown in jail as a German spy. However, he was released two months later when it was proven that he had served the Russian army as a medical officer.[24] While he had been absent, the family again moved back to Kontiniusfeld, and took up residence first with Maria's parents and later in a second house owned by great grandfather, B. B. Janz 1852 and continued to live there until the death of Maria Janz and PJ 1883's marriage to Aganetha Schartner. Among his children's strongest memories of his character, his personal piety and integrity stand out. Family devotions, morning and evening, and a strong emphasis on honesty in one's dealing with others and with each other were the keystones in the moral upbringing

[23] He was a part of a group of approximately 6000 medical workers whose costs were paid for by the Mennonite villages and colonies as part of the effort to convince the Czar and Russian society that the Mennonites were doing their duty as citizens of the state. See Epp, *Mennonite Exodus,* p. 28.

[24] HP 1913 in an interview with me in 1990. Also confirmed by Lena Rahn Pauls 1914 in conversation, that he had already left for the war before she was born on 18 August 1914. See also PJ 1910's memoirs.

of his children.[25]

The death of his mother, Maria Janz Rahn 1888, coupled with the death of his sister a month later, of course, had a profound impact on my father and his siblings. He often recounted what a loving, gentle and praying mother he had. Both he and his sister Lena told me the story of my grandmother's death. Apparently, while very ill with typhoid fever, my grandfather had approached her the evening before she died and asked her if she was leaving him [meaning, Was she going to die?]. She simply replied, "Not yet" or something to that effect and asked for a cup of water.

The next morning she called him over and said to him, "Now I am ready to die. I have now committed everyone of our children to the Lord, and can die in peace." A few hours later at 11 AM[26] on Christmas Eve she passed away. The pain of that death was, of course, renewed a month later by the death of my father's little sister, Maria 1918. Although very young she was remembered for her happy disposition and her piety.

My father often said, "She was too good to live," and recounted how Maria1918 would suddenly cease playing and go stand in a corner.

When asked what she was doing, she would say, "I am praying."

After his wife's death, on 29 Sept 1921 Peter 1883 married Maria Janz's friend, Aganetha Schartner [born 2 March 1889, died 6 July 1968] who had worked as a nurse in an orphanage for some nine years in Grossweide, a village about 10 kilometres from Kontiniusfeld. This marriage produced three more children, Justa [born 10 August 1922, died 1 July 2004], Mika [12 April 1924, died 2005] and Anni [born 11 August 1927]. Though this second marriage was entered upon under some duress, for Peter had a family of four small children, there was and still is the greatest respect for the second mother by everyone in the family. My father, though terribly lonely for his birth mother, had nothing but positive things to say about his stepmother, a reaction that is echoed by Lena Rahn Pauls 1914. As Aganetha's correspondence to my father shows, she never seemed to make any distinction between her own and her predecessor's children. While it is

[25] See Anni Rahn Wiens 1927, *Gepriesen sei die Gnade Gottes,* p. 6.

[26] JP 1910, *Memoirs,* p. 5.

also clear that Peter Jacob 1883 had first married his childhood sweetheart, and, according to Aunt Justa, never loved his second wife with the same passion, he too displays the utmost respect and loyalty to her and her children. Even today, there is till remarkably good communication between the various offspring of my father's siblings and half siblings. A deep sense of family still persists.

The second marriage also benefitted the family financially. In the same epidemic of Typhoid and Scarlet fever exacerbated by the famine of 1920/21 that led to Maria Janz Rahn's death, Aganetha Schartner's father also died.[27] After their death and shortly after PJ 1883 and Aganetha 1889 were married they purchased the Schartner house, probably from the family estate. Since both Aganetha's brothers [David 1891 and Abram 1894] already had substantial personal property in Kontiniusfeld, it seems that the house was more affordable than might otherwise have been the case. It was in this house that the family continued to live until they left Kontiniusfeld with the German army in 1943.

The other very significant event in the life of PJ 1883's family that gave rise to the correspondence published in this book was the separation of my father, Jacob Peter Rahn 1910, from the rest of his family through his emigration to Canada. Though his own account of his trip [he called it an escape] from Russia is included, the immediate events that created that separation need to be narrated.[28] Apparently PJ 1883 and ASR 1889 together with their friends, a Henry Kliewer family, had decided to emigrate from the Soviet Union in the Autumn of 1929. The plan was to send JP 1910 and grown children from the Kliewer family on ahead to Moscow to get the proper emigration papers and tickets, while the families prepared the final

[27] Anni Rahn Wiens 1927, p. 2. Aganetha's Mother, Maria Retzlaff Schartner had died in 1915. According to Lena Rahn Pauls 1914 fifty people out of a village population of approximately 550 [1908 census] died during that year in the village of Kontiniusfeld.

[28] This story has been pieced together from what my father and my Aunt Lena Rahn Pauls 1914, told me. My father also left a written account of his emigration in both English and German. See pp. 287-288. In addition other details emerged in a transcribed interview [1990] with my uncle, Henry Peter Rahn 1911 and from my Aunt Justa Rahn Prei Geddart's autobiography, p. 4. Finally, letters 1930 06 13 & 1930 07 15 give evidence of the uncompleted auction sale.

affairs and then joined them in Moscow. To that end on 8 November 1929 the Rahn family held an auction sale of most of their possessions. Toward evening my grandfather [JP 1883]asked my father [PJ 1910] if he wished to accompany Abe Kliewer [a younger brother to Henry] and a sister, Liese, and her husband, Abe Kaspar, to Moscow by the night train to set the plan in action. The final details were being taken care of at home, such as packing their bags and preparing large quantities of roasted buns [*Jereischte Tweiback*] by Aganetha Schartner Rahn and the girls. After my father left the yard, the GPU came to question my grandfather. Later upon his confession of having held an auction sale they arrested him and took him off to Halbstatt. He was held for two days and then released upon his promise not to leave the Soviet Union. Thus the family never made it to Moscow, but remained behind in Kontiniusfeld. Within a month my father and Abe Kliewer had successfully managed to get exit visas and left Moscow on 5 December 1929 by train for the Latvian border. Though the Abe Kaspars had also sought authorization to leave, somehow their exit visa applications were misplaced and they never got out of Russia. After some time in a German refugee camp my father arrived in Canada [St. John, New Brunswick] on 17 March and reached Coaldale Alberta on 22 March 1930. He never saw his parents again and remained separated from all his siblings except Aunt Justa for some forty years. Aunt Justa managed to get to Germany in 1943 and established contact with him shortly after World War II ended.

1930
IN KONTINIUSFELD

Family name abbreviations used in discussion of the letters

ASR 1889	Aganetha Schartner Rahn 1889 – 1968
PJ1883	Peter Jacob Rahn 1883 – 1942
JP1910	Jacob Peter Rahn 1910 – 1991
HP 1911	Henry Peter Rahn 1911 – 1996
JohP 1913	Johann (Hans) Peter Rahn 1913 – 1944
HelP 1914	Helena Peter (Rahn) Pauls b. 1914
JusP 1922	Justa Peter (Rahn) Prei Geddart 1922 -2004
Anni P 1927	Anna Peter (Rahn) Wiens b. 1927

Karte

1930 01 17

Wie geht es Dir?

Lieber Jakob: Da Ich endlich mal erfahren durfte wo du bist, durch Joh. K. [s] Brief, so versuche ich dir ein Lebenszeichen und Liebeszeichen zu senden. Haben deine Karte von d. Grenze erh., weiter nichts. Gesund sind wir, wünsche d. v. Herzen! Wie geht es dir? Was arbeitest du? Wie war die Aufnahme? Hast Aussicht bald G.Vater zu sehen und nach Amerika geschrieben? Was ist in den Blättern? Gibt's Erfreuliches für uns? Was kosten die Produkte? Zucker, Mehl, Kartoffeln, Fleisch, Zuendhoelzer? Feiertage, Gebetswoche und Bibelbesprechung in aller Stille verlebt. Wetter maessig. Ziehen schwere Wolken auf, sind ganz in der Nahe, – Blitze

Postcard

1930 01 17, received in Hammerstein refugee camp 27 Jan 1930

How are you doing?

Dear Jacob, Since I was finally able to ascertain where you are through the Joh. [Johann] K[liewer's][1] letter. I will try to send you a token of our being alive and of our love. Have received the postcard from the border, otherwise nothing. We are healthy, heartily wish the same for you. How are you doing? What are you working at? How was the reception? Do you expect to see grandfather[2] soon? Have you written to America? What is in the papers? Is it happy news for us? What do the groceries cost? Sugar, flour, potatoes, meat and matches? Spent the holidays, prayer week and bible study [week] in complete peace.[3] Weather is normal. Heavy clouds are approaching, are quite nearby. Lightening is

[1] Probably Kliewer, for JP 1910 was traveling with Abram Kliewer, Johann Kliewer's son.

[2] B. B. Janz 1852 who had emigrated from Kontiniusfeld to Coaldale in the 1920s.

[3] On 24 December 1929 the Soviet regime declared an end to Christmas and the birth of Jesus because according them *it had never happened*. In addition according to USSR offical statistics, at least 579 churches had been closed in the weeks leading up to Christmas. DB 19/02/1930, p. 4. This activity seems not to have reached Kontiniusfeld at this time.

zucken, werden bald auf Mara sein! ... + Ab. Regier ... Glück un. Du. Vertrauen nach oben ist Hoffnungsschimmer! Wiedersehen, man weiß nicht, Gott w. m. d. d.Sachen gehen w. e. h.ist.[i] Dein ... W. [arten] auf Antwort. [Peter Jacob Rahn 1883; PJ1883]

[i] *Gott wird's machen, dass die Sachen gehen wie es heilsam ist. – H. Woelk.*

flashing. Will soon reach Mara.[4] ____ ____ ___[5] and Ab. [Abram] Regier[6] [had] luck and you. Trusting in the one above is a gleam of hope. Aufwiedersehen, man knows not, but God will see to it that events turn out for the best.[7] Yours [masculine *Dein*]... Await your reply. [PJ 1883]

[4] Maybe a local landmark or perhaps a reference to Mara, the place of bitterness in the biblical book of Ruth.

[5] Three or four words crossed out.

[6] Partially readable on the postcard, partly crossed out.

[7] The last clause is written in single letter abbreviations. I owe the translation of the meaning to H. Woelk.

Karte

1930 02 19

Fahre nach Brasilien

Lieber Bruder! Wünsche dir die beste Gesundheit zuvor! Du schreibst ich soll mal was aus dem Dorfe schreiben, so will es versuchen. Hier gibt es nicht viel was, alles spricht nur von nach Brasilien fahren, alle sagen, wenn sie erst mal so glücklich sein könnten wie du. Hier gehen alle zu unterricht. Im Kollektiv sind wir noch nicht, so ist noch alles beim Alten. Heinrich Siebert und Johann Janzen bestell[en] sehr zu grüssen. Wenn ich mal mit dir mündlich sprechen könnte, dann würde ich schon zu erzählen wissen. Fahre nach Brasilien, ich will auch dahin. Grüssend, Heinrich [HP 1911]

Post card

1930 02 19, received 26 Feb. 1930 in Moelln refugee camp, in Germany

Go to Brazil

Dear Brother, First of all, I wish you the best of health. You write [request] that I should write something about our village. So I will try. Nothing much is happening here. Everybody is talking about going to Brazil. Everyone says, if only they could be as fortunate as you are. Here everyone is attending classes. Thus far we are not yet part of a collective [farm]. So we are still operating under the old system. Heinrich Siebert and John Janzen ask me to greet you heartily. If I could actually speak to you face to face, then I would have much to tell. Go to Brazil. I want to go there too. Regards, Henry [HP 1911]

Karte

1930 02 19
Fünfundvierzig Rubels zu achtzehn Dollar

Lieber Sohn Jakob! Deinen Brief vom 21/1 und Karte vom 6/02 sind in unserem Besitz. Haben daraus ersehen, das du noch schön gesund bist, was wir auch noch erfreuen dürfen. Du schreibst dass ihr nach Kanada wollt, wir werden nach Brasilien gehen. Mir wurde gestern Geld angeboten, 45 Rubels zu achtzehn Dollars. Die solltest du dann abzahlen. Würdest du damit einverstanden sein? Wir werden das sehr nötig brauchen. Im Kollektiv sind wir bis heute Gott sei Dank noch nicht. Will auch nichts davon wissen. Es will einem beinahe zu lange dauern, bis wir endlich mal von dem allem erlöst sind. Haben einen schönen Winter. Es ist sehr gut, das du auch von Kliewers ihre schreibst und sie von dir. Kliewers wollen auch nach Brasilien. Hier fast alle. Schreib nur mehr, wie es sieht und wie es dir geht. Nebst Gruß verbleiben wir d. fürbetende. *Heinrich will auch noch geschriebene Antwort erhalten.*[ii] [PJ 1883]

[ii] Besondere Nachschrift, geschrieben auf der linken Seite von der Karte.

Post card

1930 02 19 Received 26 Feb 1930
Forty-five rubels for eighteen dollars

Dear son Jacob! Your letter of 21/1 and card of 6/02 are in our possession. Have discovered from them that you are still in good health, that which we also still enjoy. You write that you want to go to Canada; we are going to Brazil. Yesterday I was offered 45 rubels for 18 dollars. You would then be expected to pay them off. Would you be agreeable to this? We will have great need of it. Until today, thank God, we are not in the Collective. Also, don't want to know anything about it. It is almost taking too long until we are finally freed from all this. Are having a beautiful winter. It is also very good that you write about the Kliewers relations[8] and they about you. Kliewers also want to go to Brazil. Here just about everybody! Write more about what your prospects are and how you are doing. In addition to a greeting, we remain yours, the ones praying [for you]. *Henry also wants to receive a written answer.*[9] [PJ 1883]

[8] JP 1910 was travelling with Abe Kliewer, the son of John Kliewers from Kontiniusfeld.

[9] Written as a postscript on the left hand side of the card.

Karte

1930 02 21

Alle aus unserem Dorf wohlen nach Brasilien

Lieber Sohn Jakob! Will versuchen, dir wieder ein Lebenszeichen zu senden. Durch Gottes Gnade sind wir noch alle schön gesund und wünschen dir dasselbe. Haben von dir eine Karte von der Grenze erhalten, und einen Brief den du bei Kliewers ihrem hattest beigelegt und eine Karte aus dem Lager. Berichte uns doch, wo du allerwärts gewesen, in was für Staedte. Bei uns ist es jetzt in allem ein wenig Stillstand, Brot haben wir noch, wenn's bleibt, werden wir auch noch eine Zeitlang haben. Die Kuh ist noch nicht Milch, steht aus dem Januar schon trocken. Wird sich in 14 Tage ändern, wird dann wieder besser gehen, wenn wir sie halten. Winter ist sehr maessig! Machen uns fertig zu Frühjahr. Werden nach Brasilien gehen, wohlen Alle aus unserem Dorf. Gut ist, dass ihr einer vom andern schreibt, tut es weiter. Die Geschw. bestellen zu grüssen! Anni spricht noch oft von dir, besonders wenn sie sagt wem sie liebt, nennt sie dich auch immer. Nebst Gruß verbleiben wir deine für dich Betende. [PJ 1883]

Postcard

1930 02 21, received 5 March 1930

Everyone from our village wants to go to Brazil

Dear Son Jacob, Will try to send you again a token that we are still alive. Through God's grace we are still well and healthy and wish the same for you. Have received a card from you from the border, one letter that you enclosed with the letter from the Kliewers and one card from the refugee camp. Tell us, why don't you, where all you've been, in what countries, cities (*Staedte*). With us everything has now reached a standstill. Food [*Brot*: bread] we still have. If it remains [with us, i.e., not confiscated] we will have food for some time to come. The cow is not yet producing milk, has been dry since January. That will change in 14 days, then that [situation] will improve, if we [are allowed to] keep her. The winter is very moderate. Are getting ready for Spring. Will go to Brazil, probably everyone from our village. It is a good idea that you[10] write separately about one another. Continue to do so. The relatives send their greetings. Anni[11] still speaks of you often. Especially when she recounts whom she loves, always names you. In addition to the greeting we remain yours who pray for you. [PJ 1883]

[10] JP 1910 and Abe Kliewer, with whom JP was travelling.

[11] Anni Peter (Rahn) Wiens 1927, JP 1910's sister.

Karte

1930 03 12
Endlich mal von der Freiheit erlöst werden

Lieber Jakob! Da heute wieder der Postbote kommt will ich versuchen an dich ein Lebenszeichen zu senden. Gesund sind wir hier noch alle, was wir dir auch von Herzen wünschen! Erhielten vorige Woche Mittwoch eine Karte von Dir auf welcher du schreibst das du den 27/II nach Kanada fährst, Sie war vom 24/II. Nun wir sitzen und warten wann unsere Stunde schlägt, wo auch wir endlich mal von der Freiheit erlöst werden. – Gott kann's machen! – So geht's noch immer wie es bis dahin. – Geld und Geld und alles Mögliche! Immer was Frisches! – P. P. erhielten von Ab. eine Karte von 27/II. Bei K. sind sie auch schön gesund. Deine Lieben bestellen zu grüssen! Nebst Gruß! Deine dich liebende und für dich betende. Auf Wiedersehen! [PJ1883]

Postcard

1930 03 12, received 3 April 1930
Finally to be released from [this] freedom

Dear Jacob: Since the post is coming today, I wanted to send you a sign of life. Thus far we are still in good health, which we also heartily wish for you. Last week we received a postcard from you in which you wrote that you were departing for Canada on 27 Feb. The card was dated 24 Feb. Meanwhile we sit and wait for the hour when we will finally be released from [this] freedom – God can make it happen – So far everything continues as in the past – money and more money and everything is possible – Always something new. P. P. received a card from Ab. on 27/02. At the K.[liewer] family they are all in good health. Your loved ones send greetings. In addition to a greeting, Yours, who love and pray for you. Farewell! [PJ1883]

Karte

1930 03 31

Vershickt ist noch nur...

Lieber Jakob! Will versuchen dir mal wieder ein Liebeszeichen zu senden. Das Säen haben wir beendet, nun noch setzen Kartoffeln und das Übrige. Haben bis hierher in der Natur schönes Wetter, anders das Gegenteil. Ziehen große Wolken herauf. Bei manchen alles aufgesch[rieben]. Versch. i. n. n. E.J.W. v. K., aber auf gesch. s.m. In Sp. sind 5 aufges. a. n. n. einer, J. R., die Übrigen sitzen noch.[iii]

[iii] H. Woelk hat die Kurzschrift meist ganz erraten: *Vershickt ist noch nur EJW von Kontiniusfeld, aber aufgeschrieben sind mehr. In Sparrau sind 5 aufgeschrieben, ausgesiedelt aber noch nur einer, J. R.* [Jakob Regehr?].
EJW – vielleicht ein Wiens. Die hatten ein Laden in Kontiniusfeld.

Postcard

1930 03 31, received 25 April 1930

Thus far only...has been banished

Dear Jacob! Want to try again to send you a sign of life. Have finished seeding, but the planting of the potatoes and the rest remains. Up to the present we have nice weather in nature, elsewhere the opposite. Large clouds are drawing near. For some everything has been inventoried.[12] Thus far only EJW[13] from Kontiniusfeld has been banished, [maybe *verschwunden*, disappeared] but several have been inventoried. In Sparrau 5 [families] have been inventoried, but thus far only one, J. R.,[14] has been sent [or "moved"] away.

[12] Next follows a sentence filled with abbreviations. Mr. H. Woelk, who transliterated these letters from the Gothic to Latin script, filled out the meaning of these abbreviations almost completely.

[13] There is no one on the maps that has the initials EW. J would be the fathers's name. EJW is probably part of the Johann Wiens family that owned a large granary and/or another Wiens [D.] who had a store in Kontiniusfeld. These would be the most likely targets. The only other possibility I can find is the Johann Willms family who had a *Groswirtschaft*. However, in MR 25/02/1931, p. 6 there is a note that preacher Wiens was removed from Kontiniusfeld. But this might fit better with the letter 1930 09 15.

[14] Probably Jakob Regehr whose name appears on my aunt's map which is of a slightly earlier time, but is missing on Jakob Matthies map from about 1940. Both EJW and JR were perhaps part of the 450 men who were despatched from Lictenau railway station to Siberia on April 1, 1930. DB 7/05/1930, p. 4; MR 11/06/1930, p. 7.

Brot haben wir noch. Kollektiv noch nicht. Kinder zu H.[ause]. Wird aber s. ag. Geist. vor, ist vorläufig r.[uhig][iv] Wintergetreide gut. Sommer mussten nehmen, ziemlich zu zahlen. Bei junge H. Ka.[sper] ein Kind gestorben. Gesch.[wister] D.[uecks] in S.[parrau] sind gesund. Wir auch, wünschen Euch allen dasselbe. Schreibt Karte. Könnt sehr ausführlich schreiben, erhalten sie. Briefe fast keine. Werden wir uns wieder sehen? Verbleiben deine für dich Betende. Gruß an alle die uns lieben! [PJ 1883]

[iv] *Kinder zu Hause, wird aber sehr agitiert ??? ist vorläufig ruhig.*

The rest are still sitting [at home]. Food we still have. Still not in the Kollective. Children are at home, but much agitation for ["them"????]. Quiet at the moment! Winter wheat good! Summer [wheat?] were forced to take it and pay quite a bit extra. At the young family, H. Ka.[15] one child has died. Relatives D.[16] in Sp. [Sparrau] are in good health. We are too, wish the same for all. Send cards. Write detailed matters, we receive them. Letters, almost none [arrive, are received]. Will we see each other again? Remain the ones who pray for you. Greet all who love us. [PJ 1883]

[15] Probably Heinrich Kaspar.

[16] JP 1910's aunt, Helene Rahn Dueck 1891's family. Dueck was employed as letter carrier.

COMMENTARY

This card is the first indication that the process of dekulakization has been set in motion in the Kontiniusfeld/Sparrau villages. Though there had been sporadic attempts in earlier years to identify and remove individual kulaks in many villages, both Mennonite and Ukrainian, a systematic policy had offically been approved by the Politburo and the GPU at the end of January/ early February 1930. The process[17] [at least in the Khortitza region] required that the village poor incited by party activists drew up a list of already disenfranchised, expert and kulak households who were to be immediately liquidated. It seems that the main criteria for determining who was a Kulak were whether the head of the household had in the past rented land in addition to whatever large land holdings he already possessed and whether he had employed hired workers. This list was then approved at both the village and district and, finally, regional soviets. The next step prior to being dispossessed of their property was to make an inventory of all that each household on the list possessed, usually done by an appraiser or small committee. Finally, the head of the household was arrested for deportation and the family removed from the premises by a seizure party of people made up of the village poor, village commissions and local collective accompanied by the GPU, officials of the district soviet, and members of the communist Youth league. Thus the appraising of the properties of several of the village households mentioned in the card above was the first visible indication of which villagers were on the list. Moreover, if the family had, as in the case above, disappeared it usually meant that they had been exiled to the Northern or Eastern Soviet Union rather than that they had been resettled into a recently established kulak village.

[17] Neufeldt 2009, pp. 271-278, discusses the process as it emerged in Khortitza in detail [based on Soviet archival evidence]. The process may have varied slightly in the Molotschna, but probably not significantly, since Dolot, pp. 16-17, describes a somewhat similar process occurring in his Ukrainian village in 1930.

Brief

1930 04 01

Der älteste Onkel wirds verstehen

Lieber Jacob, Weil Ich Gestern nicht geschrieben habe, so werde Ich Heute versuchen etliche Zeilen zu Papier zu bringen. Es war Gestern ein schöner Sonntag. Hatten in der Versammlung einen Besuch. Zwei Brüder dienten mit dem Wort Gottes. Der Eine sprach über Jesaja 14, der Andere über Hosea 11, die texte. Wunderschön wie Gott uns doch nicht gleich im Stich lässt, sondern Uns immer versucht herum zu holen, wie wird Er dann eine Last auferlegen, aber dann auch tragen hilft. Und wie geht es Dir? Warum willst du nicht mit N. R. Neufeld zur Versammlung fahren? Durftest du nicht mit fahren oder wolltest du nicht? Dann schreibst du das du nur ein und ein halb Monat da sein willst. Hättest du nicht besser da bleiben, damit du eine beständige Stellung hättest auch im Winter, sonst wirst im Sommer gut verdienen und im Winter ohne Stelle sein. Und das wäre dann schade. So dass es vielleicht besser wäre eine beständige Stelle zu haben. Teile dir noch mit das der Major uns abgegangen ist. Wir mussten Ihn kastrieren lassen und nun bekam

Letter

1930 04 01

The oldest Uncle will understand

Dear Jacob, Since I did not write anything yesterday, I will try to put a few sentences on paper today. It was a nice day yesterday. Had a visit in the church. Two brothers presented the Word of God. The one spoke about Isaiah 14, the other about Hosea 11. Wonderful how God does not leave us in the rut, but tries constantly to turn us in another direction, how He lays a burden upon us, but then helps us carry it as well. And how are you faring? Why don't you want to drive to church with N. R. Neufeld? Were you not allowed to go along, or didn't you want to? Then you write that you only want to stay there [at your present employment] for one and a half months. Wouldn't it be better to stay there, so that you might also have a permanent position in the winter? Otherwise you will earn well in the Summer and be without a position in the winter. And that would then be unfortunate. So it might be better to have a steady position. Am also informing you that Major [the work horse] has departed from us. We had to have him castrated,[18] and then he developed a severe

[18] Part of the communist ideology was to industrialize the agricultural process. In particular the tractor was to replace the horse. Thus regional or district commissions undertook to curtail the private breeding of horses, one of the main supports of the private peasant farmer. By reducing the availability of horses they

Er starke Krankheit. Furchtbar gelitten. Es war ein Jammer anzusehen. Und es war nur ein Tier. Und dann gibt es Menschen, die müssen noch viel mehr aushalten. Dann gibt es Menschen die haben an solchen Gemärter noch ein wohl gefallen. Im Norden sint Es – eute wollen wir den Mist verwerfen, das ist ja eine schwere Arbeit. Dann in den nächsten Tagen werden wir im Kraut arbeiten müssen. Es ist viel im Getreide, überhaupt wildes Wrukel und Tollkraut ist eine ganze Masse. Erwarten heute einen Brief von dir. Haben letzte Woche keinen erhalten. Ich muss schließen, muss auf Dorfsarbeit gehen. Das andere Mal schreibe nur recht viel. Kannst mir vielleicht eine politische Zeitschrift schicken. Verbleibe deine für Dich Betende, nebst Gruß deine Eltern und Geschwister.

Werde jetzt weiter schreiben. Haben ebend Abend brot gegessen. P.[apa] ging zu Is[aaks?]. Die Kinder wollen sich Sirop und Milch Konfekt machen. Das Wetter is bis jetzt schön gewesen. Jetzt friert es ein bisschen. Der Traktor pflügt noch. Haben heute 10 Hühner verkauft. Enten haben wir noch 8, eine ist uns verschwunden. Aber die lassen sich schmecken. Tante Kl.[iewer] ist sehr krank gewesen, ist auch noch krank.

illness. Suffered terribly. It was pathetic to see. And that was only an animal. And then there are people who even have to endure much more. And then there are people who derive pleasure from such pathetic suffering. In the north there are [such] for [these] people we want to scatter their manure pile, that is a very difficult task. So in the next few days we will have to work among the weeds. There is much in the wheat, especially wild (*Wrukel*) [oats?] and couch [or twitch] grass (*TollKraut*), a huge amount. Are awaiting a letter from you today. I have to close, have to go on village work detail. The next time I will really write a lot. Maybe you can send me a political newspaper. Remain yours, the ones who pray for you, greetings from your parents and relatives.

Will now continue writing. Have just finished eating supper. P.[apa] went to Is[aaks?]. The children [Justa, Mika & Anni] want to make candy out of syrup and milk. The weather has been nice up till now. Now we have a little frost. The tractor is still [able] to plow. Sold 10 chickens today. We still have 8 ducks; one has disappeared. My, but they are tasty! Mrs. Kl.[iewer] has been very sick, is still sick. Have not told her that

hoped to force the private farmers into the collective more quickly. Moreover, removing the horses reduced the mobility of the peasant farmers. DB 31/12/1930, p. 3.

Erzählen Ihr das garnicht das die E.[ltern] fest genommen sind. Wer weiss wie es David Dirksen ihre Geschwister mag gehen. Dann sind wir dich auch dankbar, dass wir noch unter Dach sind, wenn es auch manchmal zu wünschen bleibt. Papa kann noch reisen, und hatte noch eine kleine Hoffnung, dass es besser ausfälle als es aussah. Aber haben ja auch einen grossen Gott dem Alles möglich ist zu verhindern und der mit uns manch eine schwierigen Lage geholfen hat. Nun ich werde schliessen. P.[apa] will auch noch schrieben. Mika sagt du solst nun grosses Konwerk machen und sie wird uns zählen und dann wurde sie Striche mahlen. Und du sollst die zählen und so fiel Aepfel schicken. Die deinen. [Aganetha Schartner Rahn 1889; ASR1889]

Da M.[ama] geschrieben, so will ich Schluss machen. Wenn da erwähnt wird vom Besuch, da war der kein Andrer als von der R.[egierung]. Es wurde vorgegeben das man von den 3 buchstaben beauftrage sei Ge und Pet. zu suchen, in Warheit aber suchte man aber nach D.[ollar], von Euch erfahren auf der Post [den] wir erhalten und wo man so

her parents had been locked up. Who knows how David Dirksen's relatives may be faring? Then we are still thankful to you that we are still under the roof [in our house], even if sometimes some things are lacking. P.[apa] is still able to travel and has a small hope that things may turn out better than it appeared. But, of course, we have a great God for whom it is possible to forestall everything and who has helped us in many a difficult situation. Now I will quit. P.[apa] also wishes to write. Mika says that you are to make a large envelope and she will count us and then she will draw strokes (Striche). Then you are to count them and send us that many apples. The ones who are yours! [Aganetha Schartner Rahn 1889; ASR1889]

Since M.[ama] has already written, I will complete it. When there is mention of a visit, none other [is meant] than one from the government (R.) They pretended that they had been mandated by the three letters to look for a Ge[19] and a Pet[er], but in truth were looking for D.[ollars]. Probably discovered via the mail we received from you,

[19] Genrich, the Russian word for Henry. My father, JP 1910, in his memoirs says that *the three letters* was code to indicate a visit from the GPU, the forerunner of the KGB. However, the visitors turned out to be local communists using their supposed connection to the GPU as an authoritative pretext.

nachdachte wer das sein könnten. Also die D.[ollar] schücken hat keinen Witz. Wenn es möglich wäre da bekannt zu geben. Warte alle Tag[e] auf die Einladung vor G.[ericht]. Bin dem übergeben wegen nicht ausfüllen. Die Zukunft seher schwarz, dunkel reicht mich zu. – Boden leer, Geld keines, vertrauen nach oben ist – Gott sei Dank. Schon manches erlebt, aber es ist noch immer wie geworden. Hatten heute Morgensegen, Js 9.6. Dann wird es auch in Zukunft gehen, wenn auch im Gef.[ägnis]. Das ist das Schöne, das der Herr gegenwärtig ist. Ist schon mancher Sturm gewesen. Und wie viel noch kommen wir wüssen? – müssen dann sagen, "Mein Vater weiss wie viel ich tragen kann." Es hat sich sehr verändert im Dorf. – und es soll in der Zukunft noch mehr – Kinder sollen in der Schule essen. Manch einer der Nachbaren sind verschwunden – Bie Abr.[am's] Vater vor der Tür wird in nächster Zeit alles die Ge.[sellschaft] einderge.[?] und weg

and concluded who it might be.[20] Sending D.[ollars] is no joke. If it is possible, register them over there. Am waiting daily for an invitation to appear before the court (G.). Have been handed over to them for failure to fill out.[21] The future is very black. Dark would be sufficient for me. Storage loft empty, money none, Faith directed upwards – Thank God. Have already experienced many a thing, but it has always worked out somehow. Had morning devotions today. Is.[aiah] 9.6.[22] Then one can also manage in the future, even in jail. That is the good, that the Lord is present. There has already been many a storm. And do we know how many more are to come? Then one must say, "My Father knows how much I am able to bear." There has been great change in the village. – And there is to be more in the future – Children are supposed to eat in school. Many a neighbour has disappeared – At Abr.[am's][23] father's place in front of the door [across the street] in the next little while everything will

[20] [that it must be us].

[21] A specified levy of grain or money.

[22] *For unto us a child is born ... and the government will be upon his shoulder...*

[23] This is Abram Kliewer, the person with whom JP 1910 left the Ukraine. Perhaps a Jakob Toews family or a Bernhard Koop is being referred to here [based on village maps].

gef.[ahren]. Bie R.[eimer] hinter der Schule ist schon. Bei K. J. angef.[angen]. Es giebt Gäste – Aber Deutsche, vom grossen Fluss hinter uns. Nun genug davon. Wollen heute Stroh einfahren. Werde ein anderes Mal mehr schreiben. Es ist bei uns wie bei I Mose 18.20 heist. Wenn M.[ama] erwehnen tut Liese ihr Brief, dann verhielt sich die Sache so. Ihr bruder G. war gerade hier und der brach[te] den Brief wo ihr Bild drinnen war, so das alles seine Richtigkeit hat. Wenn du schreibst, schreib doch mehr. Oder hast du [nur] die eine Seite für uns? Es ist bald so weit, das wir es ausnutzen müssen dann.-------Verbleiben deine für dich betende. [An der Seite angeschrieben von PJ 1883.] Wenn du es nicht wirst verstehen, gib es den ältesten Onkel. Der wird es verstehen.

be handed over to the state and hauled away. At R. behind the school[24] it[25] is all finished. At K. J. it has begun.[26] There are guests[27]– but Germans, from the big river behind us. Well enough of that. Want to haul in straw today. Will write more at another time. Our situation resembles what is described in Genesis 18.20.[28] When M.[ama] mentions Liese's letter, the situation is as follows. Her brother G.[erhard] was just here and brought the letter, with her picture, so that everything would be proper. When you write, write a little more. Or do you only have one side for us? It will soon reach the stage where we will need to take advantage of it [empty side of JP 1910's letters]. Remain yours, praying for you. If you should not understand, give it to the oldest of your Uncles. He'll understand it![29]

[24] meaning on the other side of the school from where PJ 1883 lived. R., probably Peter Reimer who was connected to a kindergarten.

[25] [the process of confiscation]

[26] This could be a reversal of initials, since there seems to be no K. J. on the village maps, but PJ 1883 often wrote things like Koop's Jacob, which is the only resident I could find with both initials.

[27] Probably communist party members. Usually, in the Ukraine in the 1930s people who were forced upon a family as fellow residents were Russians.

[28] *...The outcry against Sodom and Gemorrah is very great and their sin is very grave...*

[29] Last sentence written as a postscript on the side by PJ 1883. The oldest of the uncles is B. B. Janz 1877.

[Auf ein anderes Blatt] An Grosspapa. Weil Sie nicht von Ohm Hein zu hören bekommen, will Ich bischen schreiben. Der ist in grosser Verlegenheit. Er ist besucht worden und dann [ist] manches genommen worden. Er war sehr verlegen, aber man muss sich wundern wie es sich noch schicken kann. Würde es schreiben, aber es geht nicht. Er schreibt vielleicht selber. Den Brief von Grossmama hat Er erhalten. Es sind viele besucht worden bei uns, aber bei uns garnichts zu finden. Wurden auch in einer Nacht gestört.

[Auf der anderen Seite vom selbigen Blatt] Ihre gewesenen Nachbaren sind nicht mehr unter Uns. Nun Denen tut uns Leid. Aber wir können es nicht ändern als zu Unserem nahen und Ihm alles anheim stellen. Die alte Tante P. lebt auch noch und muss diese Zeit durch machen. Mit Der haben sie es schwer.

[On a separate sheet] To Grandfather,[30] Since you do not hear much from Ohm Heinrich,[31] I want to write a little bit. He is in great perplexity (*verlegenheit*). He was visited [by the local authorities] and some things were confiscated. He was very upset, but one must wonder at how these things right themselves. I would describe it, but it is not allowed. Maybe he will write himself. He received the letter from Grandmama. Many people have been visited, but absolutely nothing [forbidden wares] was found at our place. Were also disturbed once during the night.

[On the other side of the same small sheet] Your former neighbours are no longer among us.[32] Those we pity. But we cannot change their fate, except to approach Ours [God] and in secret put everything before Him. The old Mrs. P.[enner, Peters?] is still living and has to live through this time. With her they [her relatives] have great difficulty. She lives

[30] This would be B. B. Janz 1852 and Grandmama would be Helena (Vogt-Penner) Janz 1854., JP 1910's grandparents, who were living in Coaldale Alberta.

[31] Ohm Heinrich Janz [26 October 1862 – 2 August 1934] was the brother of B. B. Janz 1852. He was an eccentric character who lived alone in the village.

[32] This is a reference to the Wiens family who owned a store in Kontiniusfeld and had lived next door to B. B. Janz 1852, prior to his emigration to Coaldale. Their exile is already mentioned in the postcard from 1930 03 31.

Die ist bei P. P. im kleinen Hauschen. Sie ist sehr ungesund alt geworden. Nun werde abrechen. Gesund sind wir Alle. Hans liest, Lena sagt der Mond scheint so helle, Lena bestellt zu gruessen. Ohm sagt er wird auch schreiben. [PJ 1883]

at P. P.[33] in the little house. She has gotten old in very poor health. Now I will bring this to an end. We are all well. Hans is reading, Lena says the moon is shining brightly and sends her greetings. Ohm [Heinrich] says he will also write. [PJ 1883]

[33] Probably Peter Penner who lived in Kontiniusfeld according to the maps.

Karte

1930 04 04

Eine Witwe mit Familie wurde ausgesiedelt

Lieber Jakob. Da wir für eine lange Zeit keine Nachricht von dir haben, so will ich dir mal ein Lebenszeichen senden. Gesund sind wir und Gott sei Dank dafür und wünschen dir dasselbe von Herzen. Gesäht haben wir schon, auch Fass Kartoffeln gesetzt, wollen noch mehr. Also bleibt noch Kreuterei und Sonnenrosen. Säen mit Onkel David[v] zusammen, jeder zwei Pferde. Haben auch noch eine Kuh, – Er auch. Geben da jetzt gut [milch]. Witterung regnerisch! Hatten in Anfangs Woche schneeiges Wetter. In S. [Sparrau] wurde eine Familie ausgesiedelt, eine Witwe. Ihr einziger Sohn wurde da verhaftet. Sollte dann mit Mutter und Schwestern weg geschickt werden, aber entging ihre Gärden. Mutter und Tochter wurden dann mit Gewalt weg geschickt nach zum Station B. Wurden dann aber von da wieder zurück geschickt veil Sohn und eine

[v] David Schartner 1891

Postcard

1930 04 04, Received 3 May 1930

A Widow with Family was resettled

Dear Jacob, Since we have received no news from you for a long time, I want again to send you a token [or sign] of life. We are in good health, thank God for that, and from our hearts wish the same for you. Have already seeded, also planted a barrel of potatoes, want to plant more. Still vegetables and sunflowers remain. Are planting together with Uncle David,[34] each with two horses. Have one cow, he as well. At the moment are producing well. Weather – rainy. At the beginning of the week we had snowy weather. In S.[parrau] a family was resettled; a widow. Her only son was arrested there. Was supposed to be sent away with mother and sisters, but he escaped from the watchmen. Mother and daughter were then by force sent away to Station L.[35] But, were then sent back again

[34] Schartner 1891, PJ's brother-in-law. At this point they were still not in the collective, although ASR 1889, HP 1911 and Johann P 1913 were working in the collective which had taken over some of the properties. However as early as Fall 1929 Mennonites were aware that the harvest of 1930 was to be surrendered to the State.

[35] Lichtenau, or perhaps the letter is "B" for Blumstein, which was near the railway and probably had a station or, perhaps, "B" is a designation such as B as opposed to "A" in which case A would probably be Klippenfeld/Stulnevo which was closer to Kontiniusfeld, and B would be Nelgovka.

Tochter fehlten. Letzte Tochter war auch verschwunden. Die fand sich dann aber und wurde nach geschickt, und jetzt aber zurück zu ihrem Heim geschickt. Wird auf der Bann [L. oder B.] abfahren, Hat im Witter so geschriehen das es bis zur Schule zu hören gewesen ist, so dass einem Dorfengelied die Augen übergegangen sind. Erwarten in diese Tage noch mehr solche Jammer. In jedem Dorf täglich wünschen davor bewahrt was bei den Nachtbahren [geschieht] [und] das schon lange. Wir sind nicht mehr zusammen. Nachtbahren übere Stufe gehen von einander. Möchten schon gerne das Lied, Himmelklänge 20, fünfte Zeile singen.[vi] Darin für dich befunden. 4/IV/30 Psalm 126. O wäre es doch bald! [von PJ 1883]

[vi] Das Lied geht so:
Wie mächtig ist die Liebe Flamme!
Sie ist viel stärker als der Tod.
Sie besucht in diesem Lamme,
Den Schöpfer selbst, den starken Gott.
Sie hat den, in dem alles lebet,
Sie hat den in das Grab gebracht,
Vor welchen Erd und Himmel bebet.
– Herr, unter wirf uns deine Macht!

because son and one daughter were missing. The youngest daughter too had disappeared. In the meantime she, however, was found and was sent on after the others, only to be sent back to her home. Will be sent off from Station L. She screamed so loudly in the open air, that she could be heard all the way to the school; so that even the eyes of the village sacristans[36] overflowed. We expect more such misery in these times. In every village we daily desire to be protected from what [transpires] at the neighbours, and that [concern] has been present for a long time. We never get together any more. Neighbours right next door [to each other] go their separate ways. Would gladly sing the song, *Himmelklaenge* 20, verse five; in addition there is a greeting for you found in it. 4/4/30.[37] Psalm 126;[38] Oh, if only it were soon. [PJ1883]

[36] sarcastic for *communist overseers* or *party members*.
[37] This is the date written in at the bottom of the page [as if to hide it].
[38] Psalm 126, 5: *They that sow in tears shall reap with joy*. This became JP's motto and really dominated his life style and outlook for much of his life here in Canada.

Karte

1930 04 27

In Sp. haben sie 4 ausgesiedelt

Lieber Jakob! Da wir von dir zwei Briefe von dort erhalten haben, wo du deine Reise beschrieben hast; so will ich versuchen dir ein Lebenszeichen zu senden. Gesund sind wir noch immer und wuenschen dir dasselbe. Hier ist jetzt gerade Stillstand, soll aber wieder in dieser Woche losgehen und zwar mit der Steuer. Soll wieder viel Ex.[vii] geben. Wer Weiß? Vielleicht auch wir mit unsere Kuh und einem Pferd. In Sp[arrau] hat es sehr toll zugegangen, da haben sie 4 ausgesiedelt: W. B., J. D., F. F. und K. S., Hier noch nur

[vii] Extraauflagen nach H. Woelk.

Post card

1930 04 27, Received in Coaldale 21 May 1930

In Sp.[arrau] they have resettled 4

Dear Jacob, Since we have already received two letters from you over there, in which you have described your journey, I also want to send you a sign of life. We are still in good health and wish the same for you. Right now things are at a standstill, but there is to be renewed activity this week based on the local tax rate. Are supposed to contribute EX.[39] Who knows, maybe also we with our one cow and one horse. In SP[arrau] events have proceeded in a mad and hectic manner. They have resettled[40] four: W. B., J. D., F. F., and K. S. Here [in Kontiniusfeld] thus far only A. K.[41] What all is going to transpire, we do not

[39] Extra taxes or goods according to H. Woelk.

[40] Or expropriated and removed from the village. In April 1930 in many villages families were driven out into the Russian Steppes with little of their possesssions. Another round of such resettlement occurred in September, 1930. Often the local villagers were forbidden to have any contact with these people or to help them in any way. MR 28/05/1930; 11/10/1930, p. 7; 3/12/1930, p. 6.

[41] WB – Bartsch, Boldt, or Braun, some of these are missing from the map of 1940 but appear on my aunt's map of earlier date. JD – Probably Johann Dueck. FF – There is a Franz Fast present in 1940, so perhaps, his grandfather, since on the earlier map, the place is shown as belonging to Johann Fast. KS – perhaps, Sukkau or Siebert. AK – perhaps Abram Kasper. He owned a brick factory.

A.K. Was noch alles kommen wird wissen wir nicht. Hoffentlich nicht viel Gutes! Man spricht von Versiedeln im Dorf und von Dorf zu Dorf. Was das Getreide anbetrifft, wird's nur eine schwache Ernte geben, Hat viel mit Erde getrieben, so dass sehr viel eingesenkt ist. Der Roggen ist unten schon gelb., weil keine Winter Feuchtigkeit ist und wenig Regen, dafür aber viel, ja sehr viel Sturm. Die Kol.-wir laufen von einander. Wer weiß was

know. Hopefully not much good![42] People in our village and from village to village are talking about moving to another settlement. With regard to the grain crop, there will only be a poor harvest. Had a lot of dust storms so that much of the crop [winter wheat] has been ruined. The rye is already yellowing at the bottom because there is no humidity in the soil from the previous winter, and very little rain and as a result much, yes, very much storm wind [dust storms]. The KOL.... [kolkhoz] we are running from one another.[43] Who knows what May 1[44] will bring? Is there hope for us? Write us down as free.[45] The letters bear the stamp,[46] but we receive them anyway. Kl[iewers] received news last week from Moelln. Waldh[eim] Rahn wrote

[42] The sarcasm here lies in the fact that the communist leadership was frequently announcing great new initiatives that would be "good" for the masses. From the writer's perspective these "goods" rarely resulted in the improvement of PJ 1883's lot. Hence the writer is only hoping that they need not endure any further "goods" at the hands of their authorities.

[43] Rather than cooperating as per the communist, or perhaps even Mennonite, ideology. It was part of the communist strategy to break down the Russian peasants' resistance to collectivization by dividing them from one another by engendering mistrust among them.

[44] A working man's holiday in Europe and the Soviet Union.

[45] On paper we are free or "call us free" but the letters are censored.

[46] Of censorship[?]

uns der 1. Mai bringen wird. – ist Hoffnung für uns? Schreib uns frei! Die Briefe tragen wohl den Stempel, aber erhalten sie dann doch. Kl.[iewers] erhielten noch vergangene Woche Nachricht von Moelln. Waldh. Rahn schreibt von unterwegs, 3 seine Töchter liegen schon, er selbst sehr geschwollen. Wird wohl der letzte Brief sein. Ist auch einer von den Unglücklichen. Nun verbleibe dein für dich Betender. Schrieb nach B.[Bruder] David,[viii] erhalte keine Antwort! Derselbe. [*Nachschrift*] Die Kliewers sind hier! [PJ 1883]

[viii] Dies ist David Jacob Rahn 1889. Er wohnte in der Kaukasus.

while underway[47] that three of his daughters are already lying down [dead??], he himself is heavily swollen. Will probably be the last letter from him. He too is one of the unlucky ones. Now I remain the one who prays for you. Wrote to B. David,[48] receive no reply. As always! [I remain] P.S. The Kliewers are here. [PJ 1883]

[47] This Waldheim Rahn is Johann Johann Rahn 1871 who lived in Waldheim and had been very active in the political and agricultural all-Mennonite Associations that tried to restore Mennonite indepedence from outside interference after the famine of 1920/21. See. John B. Toews, *The Mennonites in Russia from 1917 to 1930, Selected Documents*, p. 152, 156, 169, 177, 178, 193, 428. According to a letter from Southern Russia in the *Mennonitsche Rundschau* (11/06/1930, p. 7) he and his family were exiled by train to Siberia. The Communists forbade him to take anything with him during this long train ride, even though other families were officially allowed up to about 480 kilos of baggage.

[48] Probably David Jacob Rahn 1889, who lived in the Caucasus.

COMMENTARY

The last two letters indicate the onset of the nationwide dekulakization policy approved on 30 January 1930 by the Politburo in Moscow that was to identify through inventory three to five percent of the peasant farmer population as kulaks, then confiscate their property for the local Kolkhoz and deport them to the North or resettle them in new villages on the Steppes [usually the former in the Kontiniusfeld area]. There were approximately ninety households in Sparrau. When one compares the references to events in Sparrau with the previous accounts of matters in Kontiniusfeld, one gets the impression that whoever was in charge of the Sparrau village soviet council seemed much more eager to fulfill Moscow's bidding than the council in Kontiniusfeld. The deportation of a widow and her children who were no threat to the agricultural reorganization, as opposed to the attack on local well-to-do farmers, indicates both the zeal with which some one was pursuing the policy, as well as how inconsistent its practical application at the village level was. At the same time the indication that even the local operatives were moved by the widow's plight suggests that there was considerable pressure from beyond the village level to execute this new policy.[49] Clearly, the treatment of Johann Rahn from Waldheim indicates that the so-called dekulakization process was also being directed from a higher level against specific targets that had been active at the political level and were considered a threat to the Communist ideology.

[49] For more insight into both the policy and ist application at the local level, see Neufeldt 2009, pp. 263-266 and Dolot, pp. 1-40.

Karte

1930 05 07
Ein bisschen Tabak

Lieber Bruder! Wünsche dir zuvor die beste Gesundheit, bin auch schön gesund. Du schreibst Ich soll mal schreiben und nicht so faul sein. Ich habe keine Zeit, muss immer Mist verwerfen, habe schon auf vier Stellen geholfen. Unsern haben wir noch nicht verworfen, wollen ihn auf die Woche. Dann fragst du ob Heinrich Siebert bald Hochzeit haben wird; noch nicht gleich. Er spaziert nicht mehr. Bestellt sehr zu grüssen. Und Musik ist nicht mehr, nur Heinrich spielt schon ganz gut. So ist hier nicht viel. Bin eben von Rudnerweide zu Hause, wollte Schuhe kaufen, waren keine. Der Tabak ist sehr teuer, 6 Rubel das Pack. Schick mal ein bisschen Tabak her. Grüssend Heinrich. [HP 1911]
Schick Ihm nur nichts.[ix]

[ix] PJ 1883 hat diesen Satz auf der Karte hinzu geschrieben.

Post Card

1930 05 07 received 14 May 1930
A little Tobacco

Dear Brother, First of all, the best of health. I also am well. You request that I should write something and not be so lazy. I have no time. Constantly occupied with spreading manure. I have already helped at four different places. Haven't cleaned up our own barnyard yet. Want to do that next week. Next you ask if Henry Siebert will be getting married soon. Not right away. He no longer goes visiting. He asks that I send greetings. And now there is no more music.[50] Only Henry is playing quite well[51] So nothing much is happening here. Just returned home from Rudnerweide. Wanted to buy a pair of shoes. There were none! Tobacco is very expensive, 6 rubles per pack. Send us a little tobacco once in a while. Regards, Henry [HP 1911] Just send him nothing.[52]

[50] This may be a reference to the Communist declaration [Summer 1930] that all choral and congregational hymn singing was illegal [DB 6/08/1930 p. 4] or it may simply be a re-enforcement of what PJ 1883 wrote in the card of 1930 04 27 about the growing disunity and mistrust developing in the village.

[51] Presumably guitar or balalika or other musical instrument. It seems from some later letters [1930 06 11] that the young men of the village often spent their leisure evenings together in an open space such as the school yard where they entertained themselves with games and music. This may imply that there is no leisure time at the moment.

[52] PJ 1883 wrote this sentence along the side of the card.

Brief

1930 05 18

Lehrer Thiessen nach Gnadenfeld

Lieber Sohn Jakob! Wünsche dir zuvor die beste Gesundheit und das beste Wohlergeben in deinem Dienst. Weil wir deinen Brief vom 13 April erhalten haben und noch nicht beantwortet, so will ich heute etliche Zeilen schreiben. Papa [PJ1883] und Lena sind zur Bibelstunde gegangen. Heinrich musste mit Lehrer Thiessen nach Gnadenfeld fahren. Hans ist ins Dorf gegangen. Die Kinder spielen auf dem Hof. Es ist Heute ein schöner Tag, es hat Freitag und Gestern schön geregnet. Der Regen war hier schon sehr nötig, das Getreide litt schon darunter. Jetzt wird es auch sehr wachsen. Der Kukurus war hübsch aufgegangen. Aber weil es so trocken war fanden die Krähen dabei und machten Schaden. Die Sonnenblumen sind hübsch. Wenn der liebe Gott seinen Segen dazu wird geben, dann kann es jetzt wachsen. Haben schon was gekarrt und geschiebert. Haben auch hübschen Zuckerrohr. Ob wir es

Letter

1930 05 18

Teacher Thiessen to Gnadenfeld

Dear Son Jacob! To begin, wish you the best of health and prosperity in your employment. Since we received your letter on 13 April and have not yet answered it, I will try to write a few sentences today. P.[apa,] and Lena have gone to Bible Study. Henry had to drive to Gnadenfeld with teacher Thiessen.[53] John has gone into the village. The children are playing in the yard. Today is a beautiful day. Friday and yesterday it rained heavily. The rain was very necessary here; the grain was already ailing. Now it will grow rapidly. The maise [corn] had sprouted beautifully, but because it was so dry the crows found it and did some damage. The Sunflowers are beautiful. If the loving God gives his blessing, then all will now be able to grow. Have cultivated some and mounded it. Also have some beautiful sugar cane; will we get

[53] In May 1930 new requirements for what was to be taught in the schools were proclaimed. [DB 7/05/1930, p. 8, 9]. Gnadenfeld was the administrative centre for the Kontiniusfeld region, where many Mennonites were questioned about their commitment to collectivization, communism and their faith. Thanks to Colin Neufeldt I have a copy of the nature of the questioning that teacher J. J. Thiessen underwent during this time period. See *Other Documents*. Thiessen was removed from his teaching post and Kontiniusfeld was without a teacher for some time. Henry Penner, who had taught my father, JP 1910, endured a similar interrogation.

genießen werden? Mann fragt so und denkt so was die Zukunft uns bringen wird. Die Zeit wird nicht besser sondern immer... Ohm war nach Memrik gefahren. Jetzt mussten Hans [HansP 1913] und Hans Fransen da immer zur Nacht hin. Mit Heinrich ist er sich böse, haben keine Gemeinschaft.[x] Hier fahren die Leute nach Berdjansk nach Wäre denn hier ist nur fertiges zu kaufen, aber alles sehr teuer. Vorigen Sonntag waren Papa und ich nach Waldheim zur Silberhochzeit gefahren, zu Rakows. Es war da noch alles ganz gut. Nur man hört da so viele traurige Nachrichten. Oft sehnt man sich nach die Stunde wo dies alles ein Ende haben wird.[xi] Wie froh müsst ihr sein wenn ihr alles hört, das ihr nicht mehr [hier] seid. Oft wird gegenseitig der Mut zugesprochen, nur Geduld und Glauben, aber der will oft schwinden, wie lang hat man schon ausgeschaut. Aber es ist alles nicht von ungefähr. Nun werde abbrechen, Papa wird auch noch schreiben, man würde noch viel können aufs Papier bringen aber.. Glucken haben wir vier sitzen, 3 auf Hühnereier und eine auf Enteneier. Will kommende Woche

[x] Ohm Hein war der allein stehender Bruder von Benjamin B. Janz 1852, der Groszvater von PJ 1883.

[xi] Aganetha Schartner Rahn war 17 Februar 1889 geboren, ist 6 Juli 1968 in Bojewoji, Omsk, Süd Russland gestorben. Also hat sie noch viel erleben müssen.

to enjoy it? One questions and wonders what the future will bring us. The times are not improving but rather... Ohm[54] has driven to Memrik. Now John and John Fransen always go there for the night. He is angry with Henry, they have no communication. Here people are traveling to Berdjansk for supplies because here only ready made goods can be bought, but always very expensive. Last Sunday Father [PJ 1883] and I drove to Waldheim for the silver wedding anniversary of the Rakows. Everything was still very nice. But one hears so many sad stories. Often one longs for the hour when all this will have an end.[55] How happy you must be, when you hear all this, that you are no longer here. Often we are advised to have courage, just patience and faith, but that is often in danger of disappearing. For how long has one already had expectations! But it is all not by chance! Now I will quit. Father will also write. One could put much on paper—Luckily we have four brooding hens, 3 sitting on chicken eggs and one on duck eggs. Want to set another

[54] OHM was an uncle, an old bachelor, Henry Janz, the brother of Benjamin B. Janz 1852, and uncle of Maria Janz, the first wife of PJ 1883. He was called "Ohm Heinrich."

[55] Some 60 years later it did but by then they were all long gone or dead! Aganetha Schartner Rahn, born 17 February 1889, died 6 July 1968 in Bojewoji, Omsk, Russia.

noch eine Glucke auf Enteneier setzen. Die Kuh gibt noch schön Milch. Die Armen kommen sich noch oft durchgelassene Milch holen. Habe auch schon 6 Käse gemacht. Die kleine Ferkel kosten 20 bis 25 Rubel das Stück. Es grüsst dich in inniger Liebe deine Mutter. [ASR 1889]

brooding hen on duck eggs this week.[56] The cow is still producing lots of milk. The poor often come to get the separated skim milk. Have already made 6 cheeses. Small piglets cost 20 to 25 Rubles apiece. Greetings to you with sincerest love, Your Mother. [ASR 1889]

[56] This was extremely important for the survival of the family, since most of their grain and vegetable harvest was confiscated by the state. Thus the production of domestic fowl provided some nutrition when other food sources were gone. Domestic fowl began to be taxed in November 1930. DB 31/12/1930, p.3.

Karte

1930 05 26

Die Ernteaussichten sind hier nur schwach

Lieber Jakob! Da wir nun schon 14 Tage kein Wort nicht von dir haben und doch bestimmt glauben das du geschrieben hast und nur hier nicht durchgelassen werden, so will ich denn doch an dich schreiben, damit du doch etwas erfährst, woraus du ersehen kannst dass wir noch leben und gesund sind, was wir dir auch von Herzen wünschen. Hier wurde gestern das Mutterfest gefeiert. Es war ganz schön, aber der Schluss war am schönsten. Es Machte ein Bruder den Schluss mit Prediger K. 11,10. Die Ernteaussichten sind hier nur schwach. Es hat in letzter Zeit geregnet, aber zu spät für Winter Getreide. So ist alles beim Alten. Nur unser Hengst ist uns abgegangen, weil wir ihn schneiden lassen mussten, von

Postcard

1930 05 26

The crop outlook here is very poor

Dear Jacob, Since for fourteen days now we have not received any word from you, but certainly believe that you have written, and that it [the letters, etc.] is not being allowed through [to us] here, I want to write to you anyway, so that you might get some news of us, in order that you might realize that we are still alive and healthy, something we heartily wish for you. Here we celebrated Mother's Day yesterday. It was very nice, but the closing was the most beautiful. A certain brother, [meaning one of the congregation] ended the celebration with Proverbs 11:10[57] The crop outlook here is very poor. It has rained recently, but too late for the winter wheat. Thus everything is following the old pattern. Only our stallion left us[58], because we had to have him castrated, demanded by

[57] *When it goes well with the righteous, the city rejoices; And when the wicked perish there are shouts of gladness.*

[58] Died of infection. According to *Der Bote* 5/07/1930, p. 3, on 1 May 1930 by decree, probably of the Central Communist Party of the Ukraine or perhaps Moscow, all horses were to be confiscated and housed in common barns. Local officials were of course charged with instituting such decrees. Thus while some village Soviets seem to have allowed private individuals to keep a horse, they certainly do not seem to have tolerated the presence of a stallion that could be used for breeding purposes in the private economic system.

der Regierung befohlen. Will schließen. Heinrich bestellt zu grüssen, und die kleinen drei schlafen noch. Verbleibe dein für dich betender. Schreibe nur viel, damit wir auch etwas erhalten. [PJ 1883]

our authorities. Will close. Henry sends his greetings. The three little ones [Justa, Mika, Annie] are still sleeping. I remain the one who prays for you. Write a lot so that we at least receive some news.[59] [PJ 1883]

[59] "At least something gets through."

Karte

1930 06 02

Noch immer dankbar

Lieber Jakob! Will versuchen dir ein Lebenszeichen zu senden. Wir sind, Gott sei Dank, schön gesund und wünschen dir dasselbe. Papa sagt ich soll weiter schreiben, denn ihm flattert die Hand beim schreiben. Deinen Brief vom 14 April erhielten wir vorige Woche. Hast du unsern vom 1 Mai erhalten? In den Tagen schickten wir einen Brief ab, auch später noch mal. Sind jetzt in der Pfingstwoche, die Jungens wollen noch alles karren. Wollen heute noch Hirse säen. Mit dem Getreide geht es hier. Unser Roggen hinter dem Garten steht sehr hübsch. Wissen nicht was noch kommen kann. Es hat hier schon vor zwei Wochen geregnet, und Sonnabend hat es wieder geregnet. Es ist jetzt alles sehr grün. Haben gleich hinter dem Abtritt Kartoffeln gesetzt, die stehen auch sehr hübsch; werden schon leben bleiben wenn es [nicht] anders wird, aber es wird immer dunkler. Wir fühlen uns noch immer dankbar, wenn wir zur Ruhe gehen am Abend unter unserem Dache. Wenn Hoffnung nicht wäre, dann lebte man nicht mehr. Aber das ist die Wahrheit. Aber oft will sie sinken. Frau Derksens Katya ist hier gegenwärtig bei

Postcard

1930 06 02

Still always feel thankful

Dear Jacob, Wish to send proof that we are still alive. We are, thank God, very healthy and wish you the same. Papa says I am to complete the letter since his hand shakes too much when he is writing. Your letter of the 14th April arrived last week. Have you received ours from 1st May? Around that time we sent off a letter, and then later another. We are now in Pentecost week. The boys want to cultivate everything. Still want to seed millet today. Here the grains are growing well. The rye behind our garden is growing nicely. Don't really know what might still happen. It rained hard here about two weeks ago. Now everything is very green. Planted potatoes just below the steps (*Abtritt*). They are now standing very beautifully. We shall remain alive, if things remain [as they are], but it is always getting more clouded. We still always feel thankful, when at night we lay down to rest under our own roof. If hope disappeared, then one would no longer live. That is the honest truth. But often one's hope begins to fade. Mrs. Derksen's Katya is presently here ihnen. Haben uns schon ein Pferd gekauft, es hat uns sehr geglückt, wieder ein Braunes wie der Major. Es ist ein williges Pferd. Papa nahm meinem Bruder A.[bram] mit und der versteht es ja. Justel geht noch zur Schule. Die kleinen Beide schlafen sich gut aus, die schelten dass du nicht an sie geschrieben hattest. Zwieback werden wir auch noch zu backen, wenn wirs behalten dürfen. Wird es uns auch noch so glücken wie euch? Oder ob der Liebe Gott was anders für uns bestimmt hat. Verbleiben in Liebe. [Erste und zweite Zeilen sind von JP 1883 geschrieben, das andere von ASR 1889]

with them. Have also bought a horse. We were very lucky. Again a brown one like our Major. It is very willing [to work]. Papa took along my brother A.,[60] who has a knack in these matters. Justel is still at school.[61] The two youngest[62] are both having a good sleep. They keep complaining that you haven't written them. We will also bake extra buns (zwieback) [to store?] if we are allowed to keep them. Will we experience the same luck as you?[63] Or has our loving God determined something else for us? We remain yours in love [First two sentences written by PJ 1883, the rest by ASR 1889]

[60] Abram Schartner 1894. He was quite an adept politician, who was mayor of Kontiniusfeld on several occasions.

[61] This note indicates that J. J. Thiessen was still teaching at this time. According to the interrogation document [see pp. 289-294], the process against him ended on June 19.

[62] Mika and Anni

[63] This question suggests that perhaps the baking of *zwieback* [buns] is part of another attempt to get away from Kontiniusfeld similar to the plan in November 1929 when the family had prepared large sacks of roasted *zwieback* that were supposed to sustain them on their immigration journey. If not, it certainly speaks of a desire to escape the communist system. The baking of the buns would then be simply a means of keeping foodstuffs such as raw flour from being confiscated by using it up.

Karte

1930 06 11
Heute ist mein Geburtstag

Lieber Bruder Weil heute Montag ist und mein Geburtstag ist, so will ich versuchen ein paar Zeilen an dich zu richten. Bin schön gesund und wünsche dir dasselbe. Die Pfingsten haben wir so verlebt: Sonntag regnete es, Montag war es schön, spielten Krug bei Lehrer auf dem Hof, D. auch aber bei Richerts. Abends ist nicht viel gewesen. Nach den Feiertagen Mittwoch und Donnerstag fuhr ich mit Chajm Leder kaufen, bekam ein Rubel den Tag. Freitag Vormittag schieberten wir. Nachmittag regnete es mal wieder. Sonnabend scheiberten wir auch. Abends war ich bei Peter Penner, wurde musiziert bis Uhr 2. Sonntag hab ich gut geschlafen. Heute will ich karren fahren. Nun zum Schluss sei noch gegrüst von Lena Doerksen. Gruessend, dein Bruder Heinrich. [HP 1911]

Postcard

1930 06 11
Today is my Birthday

Dear Brother, Because today is Monday and my birthday, I will try to direct a few lines to you. Am healthy and wish the same for you. We spent Pentecost as follows. Sunday it rained. Monday was nice, played (*Krug*)[64] at teacher's on the yard. Also played D.[ominoes], but at Reicherts. In the evening not much happened. After the holidays, Wednesday and Thursday, I went off to buy leather with Chajm.[65] Received one rubel per day. Friday morning we disked [in the fields] (*schieberten*). In the afternoon it rained again. Saturday we disked again. In the evening I was at Peter Penner's place. Made music until 2:00 AM. Sunday I had a good sleep. Today I want to go cultivate (*karren*). Now to close Lena Doerksen sends you her greetings. Greetings, Your brother Henry. [HP 1911]

[64] pitchers? some kind of game!

[65] This is probably a native Ukrainian or Russian who operated some sort of business. This letter shows the enterprising spirit of not only the Rahn family, but of many of the *Anwohner* villagers to work as temporary hired hands wherever necessary to earn a living.

Brief

1930 06 13

Ich würde Dir raten

Lieber Jakob, Da es heute regnet, will ich Dir paar Zeilen der Liebe senden. Sind schön gesund und wünschen Dir dasselbe. Erhielten von Dir den 11.6. die Briefe vom 5.10. und 19.5. Ich habe mir immer gedacht, Du seiest jeden Sonntag in der Versammlung und nun bist Du noch immer zu Hause gewesen. Du solltest nur mal fragen, ob Du auch mitfahren darfst. Du schreibst, dass Du 40 Dollar den Monat erhältst und Dirksens schreiben nur von 35 Dollar. Hat Neufeld Dir die 5 Dollar zugelegt. Dann schreibst Du, dass Du zum Sommer zu den Onkeln willst. Meinst Du, Du wirst es da besser haben. Ich würde Dir raten so wenig als nur möglich mit ihnen zu tun zu haben, denn es wird auf die Länge Dir doch nicht gefallen und dann bist Du der Schlechte, so wie ich es war und auch Da.[vid] Dirksen. Ich kenne sie, aber Du hast ja deinen Willen, ich wollte Dir nur raten. Du fragst in einem Brief, was wir noch alles vom Ausrufe gehalten haben.

Letter

1930 06 13

I would advise you

Dear Jacob, Since it is raining today I will write a few lines to you. We are in good health and wish the same for you. Received your letters of 05. 10 and 05.19 on 06.11. I had always thought that every Sunday you were in church but in fact you were always at home. You should simply ask if you are allowed to go along [to church]. You write that you receive 40 dollars per month but Dirksens write only about 35 dollars. Has Neufeld added 5 dollars for you? Then you write that for the summer you want to go to the uncles.[66] Do you think it will be better for you there? I would advise you to have as little as possible to do with them, for in the long run it will not please you there, and then you will be the bad guy just like I was and also David Dirksen.[67] I know them, but you, of course, have your own will. I just wanted to give you some advice. You ask in one of the letters what we still have left over from the auction

[66] ... to work for the Janz uncles, B. B. Janz 1877 & J. B. Janz 1884

[67] ...[just as happened to PJ 1883 and David Dirksen another relative of the Janzes who later lived in Pincher Creek].

Den Major, die Kuh, die Hühner, den Pflug, den Zweiräder, die Hobelbank, 1 Legestall, 2 Tische, 4 Rohrstühle, den großen Stuhl und die Schüsselbank, die Jätkarre und zwei kleine Bänke. Der Major ist krepiert vom Schneiden, haben uns jetzt wieder solchen gekauft – 30 Rubel. Er ist nur älter, aber ich glaube besser. Und im Herbst kauften wir einen Jährling zu 33 Rubel, so dass wir jetzt wieder zwei Pferde haben und eine Kuh. Die Kuh ist aber jetzt viel besser, wie sie bis jetzt war, gibt viel Milch, hat bis 60 Pfund Milch gegeben, auch sehr dicken Schmand. Weizen haben wir so viel bekommen, dass wir werden ausreichen zu Brot und zum Säen, aber soll wieder alles weggefahren werden, wird schon Vorbereitung getroffen dazu. Roggen werden wir noch etwas verkaufen können und wenn es

sale.[68] Major [a horse], the cow, the chickens, the plow, the two wheeler [carriage], a wood working bench, a bed, 2 tables, 4 cane chairs, a wash stand, a cultivator ["*jatkarre*"] and two small benches. Major has died because of the cutting,[69] have bought another similar one – 30 rubles. He is older, but I think better. And in the autumn we bought a yearling for 33 Rubles, so that we now have two horses again and a cow. The cow is much better now than formerly, produces a lot of milk, has produced up to 60 pounds of milk, also very thick cream. We have harvested enough wheat,[70] that we should have sufficient bread [for the winter] and seed grain. But everything is supposed to be hauled away again. Preparations are already being made to that end. We should be able to sell some of

[68] According to Aunt Justa [JusP 1922, *Mein Wegbegleiter Jesus Christus* p. 4] before JP 1910 left for Moscow with Abe Kliewer, the family held an auction sale of all their possessions and intended to join JP in Moscow. They had roasted several sacks of Zwieback to sustain them on their way. However, just as JP 1910 left with Abe Kliewer and sister Liese Kliewer and her husband, Abe Kaspar, for their journey to Moscow, the GPU (*miliz*) drove onto the yard. Later PJ 1883 was arrested, incarcerated and questioned for 2 days. Upon his release, thoroughly shaken by his ordeal he gave up the plan to emigrate and attempted to recover as many of his material possessions as possible in order to continue life in the village.

[69] The local government demanded that the stallion be castrated. See 1930 05 26

[70] He must be referring to winter wheat.

so bleibt wie jetzt, dann wird im kommenden Jahr noch mancher wieder hungern, trotzdem es auf den anderen Stellen eine sehr gute Ernte geben wird, denn Fleisch ist in den kleinen Städten nicht mehr zu bekommen. Kühe, Schafe, Schweine sind nicht mehr zu haben. Man spricht, dass jetzt Kamele und Pferde an der Reihe sind. Die Eier sollen alle abgeliefert werden, auch die Milch. Aber die Leute tun das auch schon nicht mehr. Milch kostet 5 – 6 Kopeken der Liter und Eier 40 Kopeken das Zehntel. Schmand ist aber viel teurer. Das macht die jüdische Regierung, die wir haben. Gibt nur nicht alles den Großeltern zu lesen. Verbleiben Deine für Dich betende. Gib Antwort. [PJ 1883]

the rye. If things remain as they are now, then many a one will starve again in the coming year [winter], even though in other areas there is a very good harvest. For meat is no longer available in the smaller cities. Cows, sheep and pigs are not available for purchase. Rumour has it that camels and horses will be next [they will also no longer be available]. All eggs and milk are supposed to be turned in [to the local collective]. But people are no longer doing this. Milk costs 5-6 Kopeks per litre, eggs are 40 kopeks for ten. But cream is far more expensive. That is what the Jewish government that we have is doing.[71] Don't give everything to the grandparents to read. We remain yours who pray for you. Make a reply! [PJ 1883]

[71] Here PJ 1883 is referring to the local or regional communist party apparatus. There seems to have been a prevalent idea that the communists were primarily Jewish, rather than Russian. This view may have arisen because in the process of collectivization, the government frequently sent strangers from other areas into Mennonite or Ukrainian villages to supervise the process. It seems that many of these in Mennonite villages were of Jewish origin. DB 31/12/1930, p. 3.

Brief

1930 06 15

Unsere Flecken waren am reinsten

Lieber Jakob, will versuchen, etliche Zeilen an Dich zu richten, was wir schaffen, wie es alles geht kann man ja nicht auf Papier bringen. Gesund sind wir, dem Herrn sei Dank, noch alle, was wir auch Dir dort in der Ferne wünschen. Haben es immer drock. Haben schon schönen Regen gehabt. Ich schaute heute schon, wie groß die Kartoffeln sein würden und fand schon solche als Hühnereier. Es wird auch große Kartoffeln geben. Es ist draußen wieder Gewitter. Der Kumst hinter dem Garten steht prachtvoll. Haben auch sehr hübsche Sonnenblumen. Hier sind solche, die haben ihre so verkrauten lassen, dass sie es nicht mal rein machen können. Papa und die Kinder und ich fuhren letzt zur Steppe beschauen, aber unsere Flecken waren am reinsten. Die Kinder kamen eben vom Maulbeeren essen. Sind noch nicht reif, aber zum Essen schon. Haben schon 35 Hühnerküchel und 11 Entenküchel. Morgen bekommt noch eine raus. Dann sitzt noch eine auf Hühnereier und will noch eine auf Enteneier setzen. Will auch gern ein Ferkel haben. Denn nach Hinweis sieht uns dunkel. Justel hat schon ein Jahr zur Schule gegangen. Die Schule ist schon

Letter

1930 06 15

Our garden patches were the cleanest

Dear Jacob, Will try to direct a few lines to you, what we are working at; how everything is going one is not permitted to put on paper. We are all healthy, thank God, which we also wish for your over there far away. We are still very busy. Have enjoyed pleasant rains. I checked today to see how big the potatoes might grow and found some the size of chicken eggs. Certainly there will be large potatoes [this year]. Outside there is a thunderstorm again. The cabbage at the back of the garden is growing splendidly. Also have very beautiful sunflowers. Some people here have let them get so over run with weeds, that they can't even get them clean anymore. Papa and the children and I recently drove out to look at the fields, but our garden patches were the cleanest. The children just arrived from eating mulberries. Aren't completely ripe yet, but are [fairly] edible. Already have 35 chicks and 11 ducklings. Tomorrow another batch will hatch. In addition there is still another nest of chicken eggs and want to set another nest of duck eggs. Would also gladly get a piglet, for according to public notice it [the future] appears dark to us. Justel has already attended school for a year. School is already geschlossen. Mika ist immer sehr besorgt, wenn wo ein Küken schreit, dann wird sie es gleich zur Glucke treiben und so nicht draußen lassen, dass nichts gestohlen wird. Ist aber auch sehr ängstlich. Änni lebt fast nur von Milch. Die isst nur wenig, aber viel Milch, hält sich auch gut dabei. Sie macht viel Freude, sie spielt viel mit der Puppe. Heinrich und Hans sind spazieren gegangen. Lena ging zur Bibelstunde. Papa wartet noch auf O.Wiens und Tante, wollten her kommen. Abr. Siebert und P. Faßt waren nach Schlugg gefahren und A. S. war ganz befallen und P. Faßt halb. Wäre nicht noch ein Russe da gewesen, dann wären sie da geblieben. Als sie A. S. hierher geholt hatten, war er schon ganz blau gewesen. Unsere sollen kommende Woche noch Sand holen. Muss Schluss machen. David Schartners waren hier. Onkel Wiens war weggefahren. Jetzt ist es Abend geworden. Zum Schluss sei herzlich gegrüßt von den Deinen. [ASR 1889]

closed.[72] Mika is always very concerned that whenever a chick peeps we quickly chase it into the nest so that it is not left outside, lest it be stolen. [Mika] is generally very fearful. Anni lives almost wholly from milk. She eats very little, but much milk. Does well on it. She creates much joy, plays a lot with the doll. Heinrich [HP 1911] and Hans [JohP 1913] have gone visiting. Lena has gone to bible study. Papa is waiting for O.[73] Wiens and Mrs., said they wanted to come visit. A. Siebert and P. Fast had driven to Schlugg [??] and A. S. was severely attacked and P. Fast a little. If there had not been a Russian there, they would have stayed there. When they got A. S. back here he was already all blue[74] Our [men] are to haul sand in the coming week. Have to bring this to an end. David Schartners[75] were here. Mr. Wiens had driven away. Now it is evening. In closing, you are greeted heartily by yours. [ASR 1889]

[72] The teacher, J. J. Thiessen had recently been interrogated by the Gnadenfeld Communist apparatus. See 1930 05 18. That school was *already* closed implies that it was earlier than usual and may mean that Thiessen had not passed his interrogation, resulting in an early closure. This is affirmed by letter 1930 10 13 and the record of the interview [see *Other Documents*].

[73] Probably means Onkel, along with Tante, and therefore means Mr. and Mrs. Based on the second last sentence it may also be a way of informing the reader that the Wiens family has left the village for they do not seem to have come to visit.

[74] We would say, "black and blue".

[75] David Schartner 1891 was her brother who also lived in Konteniusfeld.

Brief

1930 06 22

Der erst viel hatte, hat heute nicht sein Eigenes

Lieber Sohn Jakob! Weil Heute Sonntag ist und wir deinen Brief vom 25 Mai erhalten haben so will ich versuchen noch etliche Zeilen zu schreiben. Gesund und am Leben sind wir Gott sei Dank noch bis jetzt, was wir dir auch von Herzen wünschen! Du fragst ob Heinrich noch lebt. Er lebt noch, muss Morgen nach Halbstadt. Es müssen viele hin, Steine und Grand fahren. Ein Pferd dürfen wir nicht geben. Haben schon alles zweimal gekarrt und geschubbert. Morgen sollen die Kartoffeln gekarrt und beschart werden. Es hat schon oft geregnet. Es hat auch gestern auf Stellen sehr geregnet. Hier nur sehr wenig, es ist gleich immer trocken, weil im Winter so wenig Feuchtigkeit gewesen ist. Haben schon hübsche Kartoffeln. Jetzt wollen sie noch einmal karren und Sand holen. Dann wird die Ernte auch schon vor der Tür sein. Wenn du den Brief erhalten wirst werden wir schon Roggen mähen. Mähen tun wir mit Onkel David zusammen. Jetzt fehlt uns unser Wagen und Putzmühle. Aber es würde ja alles gehen wenn es bloß anders würde. Fühlen uns noch alle Abend dankbar das wir noch dürfen unter dem Dache ruhen. Aber fragend steht

Letter

1930 06 22

Who once had much, today has nothing

Dear Son Jacob! Because today is Sunday and we have received your letter, dated 25 May, I will try to write a few sentences. Healthy and alive to the present moment we still are, thanks to God. We heartily wish the same for you. You ask whether Henry is still alive. He is still alive, has to go to Halbstad tomorrow. Many have to go there to haul stones and gravel. We are not allowed to contribute[76] a horse. Have already cultivated and mounded everything twice. Tomorrow the potatoes are to be cultivated and heaped up. It has already rained often. Even yesterday it rained hard in several places, but here in our area only very little. It always dries out very quickly because there was so little moisture in the wintertime. We already have beautiful potatoes. Now we want to cultivate once more and bring in some sand. The harvest time will already be at our doorstep; by the time you receive this we will already be mowing the rye. We are going to do the mowing together with Uncle David. We miss our wagon and cleaning sieve. (*Putzmühle*). But everything would be endurable if only things were different. Still feel thankful every evening to be resting under our own roof. But we do remain here

[76] To make use of or in place of a hand labourer.

man da was uns die Zeit bringen wird. Haben 35 Küchlein und 24 Entenkuchlein. Von 27 Eier sind 24 ausgekommen. Noch habe ich eine Glücke auf 10 Eier sitzen. Es hat sehr geglückt, zwei Glücken habe ich noch auf Hühnereier sitzen. Die Hühner wollen in diesem Jahr gar nicht brüten. Tauschten zwei Glücken in Tschernigowka ein für Hühner. Will noch setzen wenn nur noch brüten um das wir im Herbst haben. Haben uns auch schon ein Ferkel gekauft von Peter Sieberts zu 35 Rubel. Die sind sehr teuer. Die Kuh gibt noch immer schön Milch, machte Gestern noch wieder Käse. Wir gehen nicht die Milch durchlassen. Wir gingen eine Zeit, aber du kennst dem Hans Janzen. Die Leute wurden unzufrieden über ihm, mit einmal hörte er auf zu mitdurchlassen. Jetzt wollte er schon wieder, er ließ schon anfragen, aber es geht niemand mehr. Wer Milch durchdrehen will geht zu Abr. Kasper. Wir werden doch woll auch wieder anfangen. Papa und Bruder

questioning what time will bring to us. We have 34 chicks and 24 ducklings. Of 27 eggs, 24 have hatched. I still have one brooding hen sitting on 10 duck eggs. We have had a lot of luck. Two hens are still sitting on chicken eggs. However generally the chickens did not want to brood this year. Traded two brooding hens for chickens in Tschernigowka. Want to set up some more brooding nests if only the hens will continue to brood so that we have some chickens in the Fall. Have also purchased a piglet from Peter Siebert for 35 Rubles. They are very expensive. The cow continues to give a lot of milk; made some cheese again yesterday. We do not take the milk to be separated any more. We took the milk there for a while, but you know that Hans Janzen.[77] People became dissatisfied with him; all at once he refused to allow people to separate their milk any more. Now he wanted to start up his business again. He had made some enquiries but no one goes there any more. If anyone wishes to separate their milk, they go to Abr. Kasper. We will probably start again. Papa (PJ 1883)

[77] In the 1929/09/24 minutes of a meeting of The Commission concerned with the Purge of the Soviet Apparatus in the Gnadenfeld region, probably this Janzen is described as *manager of a separator office in Kontniusfeld , born in 1902, a middle peasant.* Judgement is *Consider him checked* [a conditional approval].

David waren zu den Bulgaren nach Kirschen gefahren, sie brachten auch ziemlich mit. Hatten Heute Piroschki, Gestern Warenki. Kirschen wird es nicht viel geben, aber Äpfel gibt es viel, wird gesagt. Die Wäre ist fast nicht zu bekommen. Auch sehr teuer, aber wenn wir in die Welt hineinschauen, dann müssen wir doch stil sein und danken, denn manch einer der erst viel hatte, hat heute nicht sein eigenes, wer weiß wie sich noch alles abspielen wird. Wenn das Hoffen nicht wäre, wurde es viel trauriger sein. Wir wissen ebend nicht was der liebe Gott vor hat und wir können uns nichts nehmen, als nur beten und dem Herrn vertrauen. Nun muss abrechen, die Kinder kommen immer stören. Papa und Lena sind zur Bibelstunde

and brother David went to the Bulgars[78] to get cherries. Brought back quite a few. Had piroschki today, vereniki yesterday. Cherries are not plentiful, but apples will be abundant, as rumour has it. To buy textiles is almost impossible. Also very expensive, but when one looks out into the world around then we must become very quiet and thankful. For many a one who once had much, today has nothing of his own. Who knows how everything will play out in the future? If there was no hope, it [our circumstances, life] would be much sadder. We simply do not know what our loving God intends and we have nothing to rely on except prayer and trust in the Lord. Now must bring this to an end. The children [Justa, Mika and Anni] are always interrupting. Papa [PJ 1883] and Lena have gone to Bible study.[79] The boys [Henry and John] have gone into the village. Mika is

[78] David Schartner 1891. The Bulgars [former residents of Bulgaria], fruit dealers and importers, lived South of the Molotschna. They had been given the land of the earlier nomadic inhabitants, the Nogais that had once lived throughout the entire Molotschna but had emigrated to Turkey in 1859.

[79] Pentecost is an important holiday in Europe. Communist authorities had imposed anti-religious policy by assigning work details to everyone on the Christian holidays of Easter and Christmas. DB 19/02/1930, p. 4; DB 19/03/1930, p. 4; DB 28/05/1930, p. 3. That two members of the family went to Bible Study was probably detrimental to participants by now, but the villagers still continued to engage in their religious practices.

gegangen. Die Jungens ins Dorf gegangen. Mika steht und bettelt sie will spazieren gehen und Justa will nicht, die will an dir schreiben, und Anni schlimmt auch. Und die Küchlein rufen auch schon und die Sonne neigt sich zum Abend. Sei herzlich gegruesst von deine Eltern. Heinrich hat schon 2 Karten an dich geschrieben. Ohm Heinrich kommt oft, holt sich auch mal Milch. Mit Heinrich ist er sich böse, haben keine Gemeinschaft. Wegen seine Leiter, Heinrich hatte die ohne Erlaubnis genommen und nicht abgebracht, das war auch nicht recht. Das kann er ihm nicht vergeben.[xii] [ASR 1889]

Das ich deine Adresse nicht schreibe tue ich absichtlich, um deinen Namen nicht zu nennen, werd es auch in Zukunft tun. Wollte sonst noch mehr schreiben aber Mama verbietet mir das auf ihr Brief zu schreiben. Nun dann werde ich den nächsten Brief schreiben. [PJ 1883]

[xii] Das Vorgehende hat ASR 1889 geschrieben. Die folgenden Zeilen hat PJ 1883 geschrieben.

standing here begging to go visiting and Justa doesn't want to; she wants to write to you, and Anni is also complaining. And the chicks are already cheeping and the setting sun gives way to nightfall. You are heartily greeted by your parents. Henry has already sent you two postcards. OHM Heinrich comes to visit often, sometimes gets some milk. He is angry with Henry [JP 1910's brother], they have no communication. It is about his ladder. Henry had borrowed it without OHM's permission and had not returned it on time. That wasn't proper. OHM Heinrich cannot forgive him this action. [ASR 1889][80]

That I do not write your address is done on purpose, because I do not want to mention your name. Will not do it in the future either. Wanted to write something more, but Mama [ASR 1889] forbids me to write on her letter. So then I shall write the next letter. [PJ 1883]

[80] The following lines were written in a different hand, presumably PJ 1883's. That ASR 1889 forbade PJ to write on her letter is, I think, the result of a concern to get the letter past the censors. ASR usually wrote letters that addressed their domestic life, while PJ 1883 often wrote about political developments. They were aware that not all letters were getting through as per earlier correspondence.

Karte

1930 07 04
Will gleich mähen gehen

Lieber Bruder. Haben deinen uns sehr werten Brief erhalten von Pfingsten. Wünschen dir die beste Gesundheit, welche wir uns auch erfreuen dürfen. Wir haben den Roggen auf dem Gartenflick schon gemäht. Will gleich bei Onkel David[xiii] mähen gehen. So ist hier nichts Neues. Die Leute schreiben von Paraguay das es dort sehr schön ist, so wie im Paradies. Willst du und Ab[ram] Kl[eiwer] nicht auch nach Paraguay fahren? Da Heinrich mähen fuhr, werde ich die Karte beendigen. Du schreibst Ich soll deine Adresse schreiben. Ich habe es absichtlich [nicht getan], damit ich deinen Namen nicht erwähnen brauchte. Habe es aus keiner schlimmen Absicht getan. David Isaaks kommt heute her zu Gast. Kinder mussten Gestern Sonnenblumen jäten, sind sehr viel Wölfe.[xiv] Lenchen ist sehr beleidigt von dir, das du von ihr nichts erwähnen tuest. Nebst Gruß! Deine für dich Betende. Schreib wieder! [Von HP 1911 begonnen, von PJ 1883 beendigt.]

[xiii] David Schartner 1891.

[xiv] H. Wölk schreibt, *Unter "Wölfe" verstand man eine Art Parsitengewäckse auf den Wurzeln der Sonnenblumenstauden.*

Post Card

1930 07 04 received 25 August 1930 in Coaldale, Alberta
Will mow the rye shortly

Dear Brother: Have received your very treasured letter written around Pentecost. Wish you the best of health, which we too are rejoicing in. We have already mown the rye that was planted in our garden patch. Will go to Uncle David's[81] to do the same shortly. Generally, here there is nothing new. People write from Paraguay that it is very nice there, like in a paradise. Don't you and Ab[e] K.[liewer] also want to travel to Paraguay? Since Henry has driven off to do the mowing, I will bring this card to an end. You advise me that I should write your address. I have omitted it deliberately in order to avoid writing your name. Didn't do it with any evil intent. Today, David Isaaks is coming here to visit. Yesterday, the children had to weed the sunflowers; there are very many "wolves."[82] Lenchen[83] is very hurt by your failing to make any mention of her. In addition to our greeting, yours, who pray for you. Write again! [HP 1911 and PJ 1883]

[81] David Schartner 1891 was the brother of Aganetha Schartner Rahn 1889, the second wife of PJ 1883. He lived in Kontiniusfeld. The dry conditions earlier in the Winter and Spring had led to earlier ripening of the crops.

[82] According to H. Woelk, who transliterated this card from the Gothic, "wolves" meant a type of parasitic growth that settled on the roots of sunflower plants.

[83] Aunt or Tante Lena Rahn Pauls 1914, JP 1910's full blooded sister.

Brief

1930 07 15

So ist Alles beim Alten

Lieber Bruder, Wünsche Dir die beste Gesundheit, welcher wir uns erfreuen dürfen. Haben deinen uns sehr werten Brief vom 18.6. erhalten. Es ist Regenwetter bei uns. Gestern mähten wir vor Mittag bei Onkel David[xv] Weizen. Bekamen ihn nicht ganz ab, da es anfing zu regnen, wurden ganz nass. Nachmittag fuhren wir unseren Weizen mähen, hatten ihn gerade aber, fing es wieder an zu regnen, wurden wieder ganz nass. Heute regnet es auch den Tag über, haben noch 12 ½ DZ zu mähen. Haben 6 Pferde auf dem Feld, spannen immer um, dann beschickt es auch. Dreschen wollen wir allein. Haben auch von Isaak Braun die Hälfte übernommen zu dreschen, es wird nicht viel geben. Hat allen Weizen stehen lassen, ist sehr verdürrt. Roggen haben wir 2 DZ, Gerste auch 2, Hafer nur 1, Kukurus 1, Sonnenblumen auch 1 DZ. Du schreibst, ich soll mal was aus dem Dorf schreiben. Stell du mir viel Fragen, ich werde die schon beantworten. Heinrich Siebert bestellt zu grüßen, er sagte, ihm sei alles trocken geworden. Er wolle nur in Amerika heiraten. Er ist bei Laury[?] Sparrau,

[xv] Schartner 1891.

Letter

1930 07 15

Thus everything remains as it was

Dear Brother, Wish you the best of health, something we ourselves are also enjoying. Received your highly valued letter on 18. 6. In our area the weather is rainy. Yesterday we were mowing wheat at Uncle David's.[84] Did not quite finish because it started to rain and we got soaked. After lunch we went to mow our wheat, had just started when it began to rain again, got soaked again. Today it has been raining all day, still have 12.5 DZ[85] left to mow. Have 6 horses on the field, keep trading them around, thus we make good progress. Threshing we intend to do by ourselves. Have contracted to thresh half of Isaak Braun's. It will not produce much. [He] left all the wheat standing, it is very withered. [Of] rye we have 2 DZ, barley also 2 DZ, oats just 1, corn 1, sunflowers also 1DZ. You request that I should write something concerning the village. Send me many questions, I will certainly answer them. Henry Siebert sends his greetings. He says for him everything has dried up. He only wants to marry in America.

[84] Schartner 1891, This must now be the harvesting of the wheat seeded in the Spring time, for the winter wheat had already been harvested much earlier.

[85] Dessiatines, about 13 hectares. 1 DZ = approx. 1.1 Hectares.

es geht ihm sehr gut. *Nun rauchen wir eine*[xvi] und dann weiter schreiben. Johann Janzen sagte, er würde schon geschrieben haben, er hat nur nicht Postpapier, bestellt zu grüßen. Am Sonntag wird viel Krug gespielt, Abends geht es zu Zeiten ganz gut. Da wird bis Uhr 2 in den Gärten geblieben und Dummheit getrieben, so ist alles beim alten. Heinrich Kaßper, Heinrich Töws, Johann Zirfass, Heinrich Siebert, alle 4 haben Bienen, es gibt dies Jahr sehr viel Honig. Der ist so bis 2 Rubel das Pfund. Heinrich Töws hat 8 Völker, die anderen nur jeder 2. Nun jetzt weiß ich nicht mehr zu schreiben. Du meinst unsere Kinder haben nicht Lust zum toben. Sie toben noch viel zuviel. Ich fragte Änni, was ich schreiben sollte, sie sagte nur, Änni weiß nicht. Sie sagt, wenn sie groß sein wird, dann wird sie auch schreiben. Justa ist schon 135 cm hoch, Mika 123 cm, Änni 103 cm. Sind schon große Mädchen. Du fragst, was wir vom

[xvi] Ist auf Russisch geschrieben.

He is staying with L. in Sparrau and doing very well. Now we'll stop to smoke one[86] and then continue writing. John Janzen says he would have written already, but has no letter paper. Sends his greetings. On Sundays much *krug*[87] is played, evenings are sometimes quite pleasant. One stays in the garden until 2 o'clock [AM] and perpetrates foolishness. Thus everything remains as it was. Heinrich Kaspar, Heinrich Toews, Johan Zirfass, Heinrich Siebert, all four have bees. There will be a lot of honey this year. It is worth about 2 rubels per pound. Heinrich Toews has 8 hives, the others each only 2. Now I do not know what more to write. You seem to be under the impression that our children[88] have no desire to romp around. They romp around far too much! I asked Anni what I should write. She only replied, "Anni doesn't know." She says that when she is grown up, then she will also write. Justa is already 135 cms tall, Mika 123 cms, Anni 103. They are already big girls. You ask what we retained from our

[86] Presumably some type of cigarette. Is written in Russian. Both HP 1911 and JP 1910 smoked as youths.

[87] Maybe some kind of game. Already mentioned in earlier letters. 1930 06 11.

[88] Meaning the young half-sisters, Justa, Mika and Anni.

Auszug übergehalten haben. 1 Pferd, die Kuh, den Pflug, den Zweiräder, die Jätkarre, die Stühle, 2 Tische, ein Bettgestell, die Schüsselbank und so das Grätschaft alles. Nun ich muss schließen, denn ich weiß nicht mehr zu schreiben. Schreibe Du noch, wo Du so recht bist, ob im Dorf oder auf der Farm. Wie Ihr dort den Sonntag verlebt, was Du die Woche verdienst. Grüß die Großeltern, alle Freunde und Verwandte. Grüßend verbleibe ich, Dein Bruder, Heinrich Rahn.[xvii] Du wirst ja lesen in den Zeitungen wie unsere Sache in M. ausgefallen ist. Schreibe doch. Du musst nicht vergessen, den Dorfsname zu schreiben. Wenn wir herauskommen gehen wir nach Paraguay.

[xvii] Genrich Rahn, auf Russisch geschrieben.

[attempted] departure.[89] 1 horse, the cow, the plow, the two wheeler [carriage], the wagon, the chairs, 2 tables, a bed frame, the washstand, and also most of the tools. Now I have to quit for I do not know what more to write. Make sure that you write, where you really are, whether in the village[90] or on the farm, how you spend your Sundays, what you make per week. Greet the grandparents,[91] all friends and relatives. With my greetings, I remain, Your Brother Genrich[92] Rahn
[PS] You will no doubt read in the papers how our affairs have turned out in M.[93] Do write! You must not forget to write the village name. If we get out of here we will go to Paraguay.

[89] See note 65.
[90] Probably Coaldale.
[91] B. B. (1852) & Helena (Vogt / Penner 1854) Janz who had moved to Coaldale.
[92] Written in Russian.
[93] Molotschna. The use of the writer's full name, even written in Russian, and the rather optimistic tone provide a sharp contrast to the more depressed tone of the parents. HP 1911 seems to have always made the best of what circumstances offered him.

Brief

1930 08 03

Was wir uns verdienen, ist nicht gekauft

Lieber Sohn Jakob, Will versuchen, noch mal wieder ein Lebenszeichen hören zu lassen. Gesund sind wir gegenwärtig alle. Justel und Mika waren diese Woche krank. Hier eine Krankheit. Aber die Mädchen toben schon herum. Es ist diese Woche auch sehr heiß geworden. Wir schauen auch schon sehr aus nach Regen. Es sind noch nur 14 Tage als es sehr regnete, nein doch schon 3 Wochen, aber dann regnete es sehr. Dem Kukurus und Sonnenblumen und Kürbis fehlt Regen. Es fehlt ein Regen, dann würde es viele und große Arbusen geben, die kleinen, was jetzt noch ansetzen, die werden welk. Die Maiarbüsche sind schon reif, haben schon gegessen. Zuckerrohr haben wir hübsches. Wünschen uns so Regen, dass es Arbusen gibt, das wir viel Sirup kochen können. Sonnenblumensamen wird es nicht viel geben. Da sind so viel diese Molozki aufgeschlagen und die nehmen die immer. Ich ging gestern gegen Abend mit Justel und Mika noch Scheiben holen zum Rösten. Heute wird schon sehr geknackt. Wir werden schon leben bleiben, wenn die Aussichten nicht so traurig

Letter

1930 08 03

What we can earn does not have to be bought

Dear Son Jacob, Will try to make myself heard once again as a token of our existence. At the moment we are all well. Justel and Mika were sick this week. Some kind of illness seemed to be going around here. But the girls are romping around again. It also got very hot this week. We are very much looking for rain. It is only 14 days since it rained, no, actually 3 weeks, but then it rained very hard. The maise, sunflowers and pumpkins are in need of rain. There is a need for rain [generally], for then the watermelons would multiply and grow large, the small ones that have just started growing are withering. The marianberries are already ripe, have already eaten some. Sugar cane we have in abundance. Are longing for rain so that we might get many watermelons to cook up a lot of syrup. There will not be many sunflower seeds. Many Molozki[94] have appeared and they are always taking them. Yesterday toward evening I went with Justel and Mika to get some sheaves [of sunflowers] for roasting and today there is a lot of cracking going on. We would stay alive if only our prospects did

[94] Some sort of bird?

wären. Kartoffeln haben wir große, die werden wir genug bekommen. Werden die bis zum Ausgraben nicht aufholen, was wir zu Hause gesetzt haben, da wo der Strohhaufen stand. Hatten auch schon einmal Borscht von unserem Kumst. Wenn es regnen tät, dann würde es hübschen Kumst geben. Getreide gibt es nur sehr wenig. Fuder, das geht aber ist so wenig drin. Auf den großen Felder ist kein Weizen in den Ähren, hoch im Stroh, aber die Ähren leer. Haben schon 30 Fuder abgedroschen. Werden kommende Woche noch nicht beendigen. Haben für kommende Woche noch ein Pferd mehr, dann wird es mehr beschicken. Haben jetzt auch 3 Fuder abgedroschen. Menschen sind wir mehr, uns fehlt ein Pferd. Wir dreschen Isaak Braun seines auch. Wenn es auch nicht viel gibt. Aber Zeit haben wir und was wir uns verdienen, ist nicht gekauft. Er gibt dann auch Stroh und Spreu. Du fragtest mal, ob wir mit Heizung ausreichen, ja es hat eben ausgereicht, jetzt haben wir Stroh zum Heizen. Haben schon was zur Mühle gebracht nach Wa.[ldheim?], aber wer weiß, ob wir es sehen werden und dann schmecken. Wir solten wieder nach

not appear so sad. Potatoes are large, we shall get enough of them. Won't take any up of those we planted here at home, there where the straw stack used to stand, until we dig them all up. Also already had Borscht from our cabbage once. If only it would rain, then we would have beautiful cabbage. There will be only very little wheat. The volume[95] is passable, but there is so little inside. In the big fields there is no wheat in the husks, lots of straw but the husks are empty. Have already harvested 30 cartloads. Will not finish next week. Have managed to get one more horse for the coming week.[96] Then it will speed up. Just finished threshing 3 cartloads. People we are aplenty, but we need a horse. We will harvest Isaak Braun's crop as well, even if it does not produce much. For we have the time and what we can earn does not have to be bought. Moreover, straw and chaff are also produced. You asked once whether we had sufficient [material] for heating. Yes, we had just barely enough. Now we have straw for heating. Have recently brought some [wheat] to the mill in Waldheim, but who knows whether we'll see it again and then taste it.

[95] A *fuder* is actually a cartload, but here seems to be used as a synonym for volume.

[96] Since the family is not in the Kolkhoz yet, they seem to have been operating similar to an Artel. They were allowed a horse and a cow while providing labour for state projects. In periods such as harvest or seed time, they were allowed to borrow extra horses from the state.

Heinrich fahren. Wenn denn wenigstens wäre darauf gab, aber wird gesagt, nein. Kauften uns wenig. Waren 2 Pfund Äpfel 5 Rubel das Pfund. Haben zwei mal Paroschki mit Mos gehabt. Hatten schon über 70 Küchel, aber dran sind wenig.(stens) noch nur 60 davon. Die kleinen sind [tod] gefallen. Habe noch eine Glucke gesetzt. Entenkückel haben wir 29 Stück, 24 sind schon große. Das hat sehr geglückt damit, keine sind zu Tode gekommen. Aron Dirksen haben 12 Entenkückel, jetzt sind schon nur 7 davon. Aber ist auch sehr aufzupassen. Aber wir würden gerne alles dem geben, wenn wir das Glück bekämen, was Du hast. Lena war zum Begräbnis gegangen. Tante Kröker, Sp.[arrau], wurde begraben, wurde am Bruch operiert und ist nachher gestorben. Ein sehr trauriges Begräbnis gewesen. Nun muss schließen, der Bogen ist voll. Sonntag ist Justel ihr Geburtstag und Montag Änni, den 10. und 11. August. Morgen wird es 3 Wochen, dass wir von Dir keinen Brief erhalten haben. Bekommst du unsere nicht? Manches würden wir noch schreiben, aber man tut es nicht. Übrigens wisst ihr alles! [ASR 1889]

We were supposed to drive over to Henry[97] again. If only there were some goods that could be obtained in exchange, but no, bought very little. 2 pounds of apples were 5 rubels per pound. Have had paroschke with moos twice. Had over 70 chicks, but only about 60 still survive. The small ones died. Have set one brooding hen on a nest. Of ducklings we have 29. 24 are already full grown. There was a lot of luck with them.[98] None died. Aron Dirksens had 12 ducklings, now only 7 still survive. For it is very needful to be watchful. But we would gladly surrender everything, if we had the luck that you had. Lena went to a funeral. Mrs. Kroeker (Sp.)[arrau] was buried, had been operated for a hernia and died afterward. Was a very sad funeral. Must close, the sheet is full. Sunday is Justel's birthday, and Monday, Anni's, the 10th and 11th of August. Tomorrow it will be 3 weeks that we have received no letter from you. Don't you receive ours? There is much we might still write, but one does not do so. But you know everything anyway. [ASR 1889]

[97] Maybe HP 1911. An earlier letter, 1930 06 22, indicated that he was supposed to haul gravel and stones in Halbstadt during the Pentecost holiday. Although according to his own last letter, 1930 07 15, he had obviously returned from Halbstadt, he probably was away in another village at this time.

[98] Farmers were counting on their domestic fowl to help them through the winter, since they had not yet been officailly declared subject to confiscation as were the grain crops. In November an official tax on chickens was introduced. DB 31/12/1930, p. 3.

Brief

1930 09 15

Diese acht Heute aufgeschrieben, und zwar alles

Lieber Sohn Jakob, weil schon eine Zeit verflossen ist, dass wir nicht geschrieben haben, so will ich mal ernst machen und schreiben. Der Wille ist immer aber einmal ist nicht Zeit, ein anderes Mal fehlt die Lust, dann sind wir in unserem Hause auch nicht sehr gesund. Papa hat schon lange immer Durchfall, das wankt hier sehr. Fühlt sich oft nicht gesund und ich habe schon über eine Woche Nervenreißen. Es hört dann mal auf und schmerzt es auch wieder dass ich nicht weiß, wo ich hin soll. Habe schon eingenommen und geschmiert und gewärmt. Es ist dann, als wenn alles nicht hilft, als nur das reine Beten. Und Justel liegt an Maseln, hat sehr den Husten dabei. Jetzt muss sie Honig einnehmen. Mika und Änni sind noch sehr munter, haben noch sehr getobt. Änni sagt, sie kriegt nun einmal Maseln. Haben heute so gelacht über Änni, sie will auch immer Honig haben und sie soll nicht. Mit einmal sagt sie, ich kann nicht pischen,

Letter

1930 09 15

These Eight Today have been Inventoried

Dear Son Jacob, Since some time has passed during which we have not written, I will now make a serious attempt to write. The will to write is always present, but at one moment there is no time, at another there is no enthusiasm in doing it. Moreover, in our house we are not very well. Papa has constantly had diarrhoea for a long time. That is very prevalent here. Often he does not feel well and I have had a nervous disease.[99] Once in a while it quits and then the pain begins again so that I don't know where to turn. Have already taken something and applied [ointment] and warmed it. At those moments it is as if nothing helps, except simple prayer. And Justel has the measles accompanied by much coughing. Now she has to ingest honey. Mika and Anni are still very lively, have romped about a great deal. Anni says she is now getting the measles. Today we laughed so hard at her, she always wants honey and she is not supposed to [have it]. All of a sudden she says, *"I can't urinate, must*

[99] The diarrhoea and the (*Nervenreizung*) shingles, eczema [?] probably reflect the extreme stress that they were feeling at this time as they saw numerous village members being evicted and their goods confiscated. See the end of this letter.

eck well mol Honig habe.[xviii] Dann haben wir so gelacht, sie meinte dann würde sie auch Honig bekommen. Jetzt neckt Heinrich sie immer, aber das macht ihr nichts. Sie spricht noch oft von Jasch. Wenn wir dann fragen, wo Du bist, dann sagt sie "Aumelikau". Als ihr Geburtstag war, sagte sie immer *Eck die jauma welt*. Die sind fast immer mit zur Steppe gefahren auf Arbusen, auf Sonnenblumen. Dann meinten sie müssten mit. Arbusen haben wir noch viel in der Scheune liegen, auch sehr schöne und viel schöne Melonen. Es wird gar nicht so viel gegessen. Melonen geben wir der Kuh. Die gibt ganz schön Milch. Sonnenblumen Samen haben wir nicht so viel wie im vorigen Jahr. Kürbisse haben wir 5 Fuder bekommen. Kukurus wird es nur wenig geben, den haben wir auch nicht zu Hause. Haben heute angefangen zu säen, säen mit Johan Wiebe zusammen. Kartoffeln haben wir ganz schöne und große bekommen. 4 Mal sind wir schon zur Station gefahren mit Getreide für uns und einmal für Isaak B[raun]. Der Regen versäumt sehr viel im Tag. Es war 2 Tage schön dann Regen. Es

[xviii] Platdeutsch.

also have honey."[100] We laughed uproariously, for she thought then she would also get honey. Now Heinrich [HP 1911] is always teasing her, but that does not bother her. She speaks often of "Jasch" [JP 1910]. When we ask where you are, then she says, "Aumelikau." On her birthday, she kept saying, *"Eck die jauma Welt."*[101] They [the young girls] almost always accompany [us] when one drives out to the field for watermelons, or Sunflowers. Then they always think they have to go along. We already have plentiful watermelons stored in the shed, very nice ones and also many other melons, nice ones.[102] It is impossible to eat so many, We are feeding the melons to the cow. She is producing plentiful milk. Sunflower seeds are not as plentiful as last year. We have harvested 5 cartloads (*Fuder*) of pumpkins. There is very little maize [corn]. We have none at home. Have begun to seed today.[103] Are seeding together with Johan Wiebe. We have harvested very nice big potatoes. 4 times we have delivered grain to the station for us [our produce] and once for Isaak B[raun]. The day['s work] is seriously interrupted by the rain. It was

[100] Written in low German.

[101] Low German expression meaning something like "Alas, what a sad world."

[102] It appears that the writer's wish for rain expressed in the last letter [1930 08 03] has been granted.

[103] Probably winter wheat. There is much talk in other letters of how the winter wheat develops or produces.

liegt noch viel Getreide auf der Steppe, sehr verwachsen. Bei unserem Nachbarn sieht es auch auf dem Hof. wird Spreu verkommen. Und dann hatte er 2 Sorten Gerste und den Hafer gedroschen und nicht geputzt. Jetzt fing es an zu regnen, jetzt hat es da aber gewachsen, das war großartig. Es ist ein sehr drocker Herbst. Das macht der viele Regen. Aber unsere würden mit allem zufrieden sein, wenn es nicht so wäre. Es sind gegenwärtig viele Leute in sehr großer Verlegenheit. Man steht fragend da, was die Zukunft bringen wird, denn es wird gefordert ohne Ende. Aber wir müssen uns fügen, der liebe Gott lässt es noch immer zu. Wie glücklich, denen das Los im Herbst getroffen hat. Wer weiß wann für uns die Erlösungsstunde schlagen wird. Die Kinder schlafen schon. Justel sagt, ich soll dir schreiben, du sollst ihr einen größeren Brief schreiben und nicht so wenig. Sie ging schon wieder zur Schule. Bruder Abram seine Käthe liegt auch an Typhus. Peter Koch[??] weilt auch nicht mehr hier auf Erden. Den wird niemand mehr antasten. Die arme Frau

nice for 2 days, then rain. Much grain is still lying in the fields (Steppe), very badly sprouted.[104] At our neighbour's one even sees it on the yard. Even the chaff is rotting. And, in addition, he had thrashed 2 types of barley and his oats, but had not winnowed it. Then it began to rain. Has it ever grown, that was astounding [to see]. It is a very busy Autumn. That is caused by the rain. But we (*unsere*) [our people] would be satisfied with everything if only things were not as they are. Presently many people find themselves in great difficulty. One stands here questioning what the future will hold, for demands[105] are being made without end. But we must fall into line, for our loving God is still allowing it to happen. Who knows when for us the hour of rescue will strike. The children are asleep already. Justel says that you are to write her a longer letter and not such a short one. She is again going to school. Brother Abram's [Schartner 1894] Kathy is in bed with typhoid (*Typhus*). Peter Koch also no longer whiles away [time] here on Earth. Him no one will bother any more. But the

[104] This early effort at collectivization was attended by many failures. Aside from the difficulty of getting sufficient labour to bring in the harvest, what was gathered from the fields often lay rotting in the open air. DB 3/12/1930.

[105] Taxation, goods in kind, etc. The communists had introduced a house or perhaps head tax to be paid in addition to the surrender of all the harvest.

mit ihren Kinderchen. Ich würde noch viel schreiben, wenn es ginge. Die Deinigen. Nebst Gruß. [ASR1889]

D. W., H. S., K. R., P. R., D. B., Witt. M. W., Wittwe H. D., K. Toe. Diesen 8 ist über Vermögen auferlegt worden, heute aufgeschrieben und zwar alles. Wie man spricht nach diesen noch 15. Ich werde nächstens alles ausführlich beschreiben, wenn noch mehr passiert ist. Derselbe. [PJ 1883 Zu geschrieben auf Aganethas Brief]

poor wife with her small children. I would write much more if only it was allowed. Yours, plus a greeting. [ASR1889]

D. W., H. S., K. R., P. R., D. B., Widow M. W., Widow (Wittwe) H. D., K. Toe.[106] These 8 had been assessed for their wealth, today been inventoried, and to be precise, everything [has been listed]. According to rumour, after these another 15. I will soon describe everything in detail when more has occurred. The same! [PJ 1883, added to the bottom of ASR 1889's letter.]

[106] From the village maps it seems that among those whose property was inventoried on this day there may have been:
P. R. – Not Peter Rahn, because he and his family continued to live in their own place with their own possessions for several years after this, but more likely Peter Reimer who had a large farm (Grosswirtschaft).
D. W. – Probably D. Wiens, the store owners.
H. S. – Heinrich Siebert, there are either two of these or one who owned two houses. The latter is probably more likely since my aunt's map only lists him once. She is probably listing the people who lived in the various houses rather than the owners. Moreover, as the owner of more than one house, he was most likely to draw the attention of the communist authorities early on in the process of collectivization and kulakization.
K. R. – … Reimer or Regehr, Not Rahn, because no K. Rahn, The K. is probably Kornelius or ??
D. B. – David Born or David Boschman
WIDOW M. W. – … Wiebe or Willms.
WIDOW H. D. – either Heinrich Duerksen or Heinrich Dueck.
K. Toe[ws] – Kornelius Toews

COMMENTARY

This new initiative to appraise the households of some of the villagers of Kontiniusfeld indicates that a brief hiatus in the dekulakization process throughout the Soviet union that had occurred in late spring and early summer 1930 was now at an end. After the publication of a letter by Stalin in *Pravda* on 2 March 1930 entitled *Dizzy with Success* that blamed the district and village soviets for some of the apparent excesses of the initial processes, the district soviets had stopped their proceedings while many soviet officials either resigned or were removed from their positions and replaced by new, usually non-Mennonite, party members. Now in late August/early September the process was reignited with increased vigour.[107] As before the first public step in the process was to make an inventory of what each of the intended victimns possessed.

The farmers [*kulaks*] referred to in this letter were the rich or Grosswirtschaften in the village and had been misinformed and used by the Communists to enlarge and enhance the returns produced by the new collectivization effort in 1930. Already in the Autmn of 1929 it had been announced that the winter wheat crop in its entirety whether privately or publicly sown then would be surrendered at harvest in 1930. However, the local authorities not only confiscated the entire harvest of 1930, but added significant taxes and levies to drive the farmers into bankruptcy. Then since they were unable to pay what was required, they were given 3 days to surrender their goods and property to the Kolkhoz or face the dekulakization process. This procedure seems to have been widespread throughout the Ukraine.[108] At this point the Rahn family remained in their home because they had relatively little property beyond their house and garden, one cow and one horse. What they did have they obtained in the 1920s after the revolution. Prior to that time they had been *anwohner* and lived in rented housing. They were less likely than many others to be designated *kulaks* at this time since they had generally worked for others rather than hired labourers themselves.

[107] For a more detailed discussion of this brief hiatus in the dekulakization process see Neufeldt 2009, pp. 278-282.

[108] DB 8/10/1930, p. 3.; MR 11/10/1930, p. 7.

Brief

1930 10 13

Es sind hier Gestern Viele ganz arm geworden

Lieber Jakob, Will vornehmen etliche Zeilen an Dich zu richten. Erhielten in dieser Woche wieder einen Brief von Dir, in dem Du schreibst wie viel Du verdienst und wie viel Du schon abgezahlt. Freuen uns, dass es Dir gut geht. O.[hm] H.[ein] lebt noch nach alter Art. Wollte nach Mamrik fahren, sehen ob die hier bleiben würden, wollte dann sein Haus verkaufen und hinüber ziehen. Aber er sagte jetzt, will ihm niemand das Haus abkaufen und so ist er bis jetzt noch immer hier. Geht dann ins Dorf und holt sich Schleudermilch. War gestern Abend bei uns, ließ sich die Kanne voll gießen, sie war nicht ganz voll. Kuh haben wir heute noch eine, aber ob wir sie Morgen auch noch haben, weiß ich nicht. Bei uns ist mal wieder die Hölle losgelassen, verlangt wird mehr wie gewachsen ist. Erst treiben sie die ins K.[olkhoz] und dann müssen die anderen herhalten, weil doch des öfteren ihr Getreide

Letter

1930 10 13

Yesterday Many Became Very Poor Here

Dear Jacob: Want to direct a few lines to you. Received another letter from you this week, in which you write how much you earn and how much you have already paid off.[109] Rejoice that it is going well for you. Uncle H.[110] still lives according to his usual fashion. Wanted to go to Mamrik to see if they[111] would stay here.[112] Intended to sell his house and move over there. But now he says no one wants to buy his house and so until now he is still here. Comes into the village to get skim milk. Was at our place yesterday evening, allowed us to fill up his [milk] can, though it wasn't quite full. Today we still have one cow, but whether we'll still have her tomorrow, I do not know.[113] Once again among us hell has been unleashed. One demands more than has been grown. First they force some into the K.,[114] and then the others[115] are the ones who have to suffer, for more often than not their

[109] Of travel debt.

[110] Probably OHM Hein, Heinrich Janz, the brother of BB Janz 1852.

[111] Presumably some of his relatives.

[112] Meaning in Russia.

[113] Clearly this letter was written before the following one without a day specified. In it the cow and the horses have been confiscated.

[114] *Kohlkoz*, collective.

[115] Meaning those not in the Collective.

verwachsen ist und in Russland muss ja alles hinaus, drum fasst man nach unserem Getreide, das im Trockenen eingebracht ist und weil nicht so viel ist wie verlangt wird, so muss man ja diejenigen aufschreiben und verkaufen, die es nicht ausfüllen können.[xix] Will versuchen zu schreiben. Es scheint, als ob nicht mal Zeit ist paar Zeilen zu schreiben. Es ist immer sehr drock. Haben es in letzter Zeit sehr drock gehabt. Mit Säen und den Boden leer fahren. Fuhren heute auch wieder mit Hafer, G.[erste] und Samen. Die Lage ist sehr traurig. Wir schauen so dunkel in die Zukunft wie noch nie. Es scheint, als ob keine Hilfe für uns ist. Es sind hier Gestern viele ganz arm geworden. Es war herzzerreißend. Ich war bei Großpapas Nachbarn zugegen. Die Mädchen haben sehr geweint. Man muss schweigen. Abends auf dem Heizen ist viel. Will schreiben was dem Papier nicht schadet. Vor 3 Wochen am Sonntage wäre bald ein sehr großes Unglück passiert. Wir waren gerade auf der Straße gegangen

[xix] Was folgt ist in eine andere Handschrift.

grain is spoiled. But in Russia everything has to be exported, so they reach for our grain that was brought in during dry [weather], and since there is not as much harvested as was demanded, one must inventory and sell [the goods] of those who are unable to fulfill what is demanded.[116]

Want to try to write. It seems as if there is not even time to write a few lines. It is always so busy. Have had a very busy time of late. With sowing and hauling until the loft is empty! Hauled oats, barley and seed [grain] away again today. The situation is very miserable. We look into the future so dark as never before. It appears as though there is no help for us. Yesterday many became very poor here. It was heart rending. I was in Grandfather's neighbourhood.[117] The girls cried bitterly. One must be silent! Evenings around the heating there is much [to talk about]. Want to write what does not harm the paper. 3 weeks ago on a Sunday a terrible accident almost took place. We had just set off down the street to W.[118] All of a sudden

[116] Different handwriting follows.

[117] Maybe in Sparrau, though more probably in Kontiniusfeld, since "Grandfather" usually refers to BB Janz 1852 who no longer lived there, but whose house was in the same neighbourhood as that of PJ 1883. Hence by referring to Grandfather's neighbourhood, the writer is actually saying, "in our neighbourhood".

nach W.[iens]. Da mit einmal schrien die Kinder auf der Straße sehr – Feuer – Dann schauen wir uns um und sehen, dass sie nach uns hinzeigen, dann liefen wir und sahen, dass es dem Nachbar sein Strohhaufen war und der hatte ziemlich Stroh. Aber weil die Menschen alle zu Hausen waren, waren gleich viele hier und die Spritzten auch und Heinrich hatte gleich Wasser genommen und nach dem Strohhaufen und weil sie unser Stroh nass machten und das Dach Pfannen hatte, ging es nicht weiter. Denn der Wind kam gerade nach unserem Hause. Wenn es bei uns noch gebrannt hätte, dann wäre das Dorf abgebrannt. Das war schrecklich. Aber P.[apa] hat auch auf gespritzt. Und dann wurde das Stroh brennend zur Steppe gefahren. Er kann es sich nach Hause holen. Es liegt auf unserem Kartoffelstück. Der Mensch hat noch keine Heizung zu Hause. Das Feuer hatte der kleine Heinz ausgelegt, nämlich mit Feuer gespielt. Ihr Mann ist begraben, der Vater liegt hier auch krank, die Tochter pflegt ihn. Gegenwärtig sind sie beide hier. Von den Verwandten ist Freitag Hochzeit, wird sehr fertig gemacht. Mit Tante wird es auch immer weniger. Hier haben mehrere Typhus gehabt.

the children began to shout loudly, "Fire!" We turned around and saw that they were pointing to our house. Then we ran and saw that it was the straw of the neighbour who had a very large stack. However, because everyone was at home [because it was Sunday], many people arrived quickly as well as the fire brigade. Heinrich [HP 1911] had immediately taken water to the straw stack, and because they watered down our straw stack and our roof was tiled, the fire did not spread any further. For the wind blew directly toward our house. If our house had caught fire, then the village would have burned down. But P.[apa] had also sprinkled. And then the burning straw was hauled out into the fields. He can haul it back to his house, for it lies on our potato patch. Now that man has no heating at home. The conflagration was started by the little Heinz, who had been playing with fire. Her[119] husband is buried, her father is lying here sick, the daughter is tending him. At the moment they are both here. Among the relatives there is a wedding on Friday, a lot of preparations are being made. For Auntie everything is deteriorating. Here several have had typhoid. P. K.[120] has been buried. May

[118] Probably Wiens, who lived nearby. They also had a friendship with a Johann Wiebe family. It is less likely to have been a Willms family since there is no mention of them as co-workers or friends in any of the other letters.

[119] Probably meaning the widow Derksen.

P.K. ist begraben, wirst das vielleicht schon gehört haben. Auch in unserem Hause sind die Kinder krank gewesen. Erst bekam Justel die Maseln, die war ziemlich krank und als die auf war, bekamen Mika und Änni. Mika hat es nicht viel geschadet, das ist wieder der alte Peter, aber Änni war sehr krank. Wir dachten eine Nacht, sie würde nicht den Morgen erreichen, hat sehr abgenommen, aß auch lange nicht, aber jetzt fing sie schon viel an zu essen und hat sich heute wahrscheinlich das Mauhchen verdorben. Sie aß heute so viel Fleisch und Borscht. Sie ist heute Abend krank, hat sich schon 3 mal erbrochen und liegt sehr unruhig. Wenn man in die Zukunft hinein schaut, dann kommt einem so viel, ach wären die Kleinen doch aufgehoben. Justa geht zur Schule. Haben nun einen Lehrer.[xx] Haben eine Woche mit dem Zuckerrohr gearbeitet. Bekammen 30 große Eimer Suppe Freitag auf Sonnabend. Nachts fingen unsere an zu pressen und Sonnabend kochten wir. Gestern kochten wir noch von 6 große Eimer Suppe von Arbusen, sehr feiner Sirup. Von Arbusen haben wir 49 Eimer Suppe gekocht. Haben noch Arbusen. Maulbeeren sind viele

[xx] Lehrer Thiessen war im Juni/1930 beiseitigt nach seine Untersuchung in Gnadenfeld. Schaut auf Karte 1930 06 02 und Briefe 1930 05 18 und 1930 06 18.

have heard that already. In our house too the children have been sick. First Justel caught the measles – she was quite sick – then Mika and Anni. They didn't harm Mika much (that is the old Peter),[121] but Anni was very sick. One night we thought she would not survive until morning. Has lost a lot of weight, didn't eat anything for a long time, but today she began to eat heartily, and has probably over eaten and harmed her little tummy. Today she ate so much meat and borscht. She is ill tonight and has already vomited 3 times and sleeps very restlessly. When one looks into the future, so much overwhelms; if only the little ones were transfigured (*aufgehoben*). Justa is going to school. We have a teacher again.[122] Spent a week working with the sugar cane. Produced 30 large pails of liquid syrup (*Suppe*) Friday through Saturday evening. [Friday] night ours [we] began to press [the cane] and Saturday we boiled [the sap]. Yesterday we also boiled 6 large pails of watermelon juice, very fine syrup. From the watermelons we obtained 49 pails of juice. Still have some watermelons. Mulberries have many rotted ones. Our Hans [JohP 1913]

[120] Probably either Peter Kaspar or Peter Kliewer.

[121] [in her]. By Peter is meant ASR 1889's husband and Mika's father, PJ 1883.

[122] This suggests that teacher J. J. Thiessen had been removed after his interrogation in Gnadenfeld. See letters 1930 05 18; 06 15; 10 13 and the postcard 1930 06 02. Also the interrogation document under Other Documents.

verfault. Unser Hans lebt fast von Arbusen. Wenn keiner am Tisch Arbusen isst, Er isst Arbusen. Schade, fangen sehr an zu faulen. Morgen wollen wir Enten schlachten. Die lassen sich schon essen. Haben schon einmal gehabt. Haben ein hübsches Schweinchen, aber glauben gar nicht, dass wir es genießen werden. Oft fragen wir uns, warum wir nicht hinüber kamen. Mehl ist keins zu bekommen, denn du kannst nicht so viel Getreide stellen, dass du darauf nehmen kannst. Nun ich werde abbrechen. Nachbars Liese ihr Töchterchen wurde kürzlich begraben, der ihre Schw. [ieger] Eltern haben auch nichts mehr. Änni liegt sehr unruhig, muss schließen. Justa wollte auch noch schreiben, aber es wurde zu spät. Zum Schluss sei herzlich gegrüßt von uns allen. Erhielten gestern auch einen Brief. [PJ 1883 und ASR 1889]

lives almost exclusively from watermelons. Even when nobody is eating watermelon at the table, he is [still] eating. Too bad, they [watermelons] are quickly beginning to rot. Tomorrow we want to butcher the ducks. They will provide a tasty meal. Have already eaten one. Also have a beautiful pig, but do not believe that we will be allowed to enjoy it [as food] (*geniessen*). Often we ask ourselves, why were we not able to cross over.[123] Flour is impossible to obtain, since one cannot steal sufficient grain[124] to bring a portion for milling [for personal use].[125] Well, I shall break off. Neighbor[126] Liese's little daughter was recently buried. Her [Liese's] parents-in-law also have nothing anymore. Anni is very disturbed in her sleep, must quit. Justa also wanted to write, but it got too late. To close, hearty greetings to you. Also received a letter [from you] yesterday.

[First part was written by PJ 1883 and the rest by ASR 1889.]

[123] To Canada. As per several earlier letters, the family had tried to sell their possessions and leave. See note 74. This is the first letter where resignation seems to be setting in that they will not be able to leave.

[124] [hold back from the authorities.]

[125] None of the flour that would be milled would be returned to those who deposited it at the mill. It would all be confiscated as state revenue. At this point the family's possessions and at least some land are still their own.

[126] The Kornelius Dirksens lived next door to the West.

Brief

Oktober 1930[xxi]

Wir hätten es besser gehabt als vorigen Winter

Lieber Jakob, Da wir diese Woche zwei Briefe erhalten haben, so will ich versuchen, sie zu beantworten. Schickten Gestern einen Brief ab. Erhielten Gestern den Brief, was mit H. seinem zusammen war. Es ist heute Sonntag. Aber P.[apa] ist nicht zu Hause, fuhr mit dem B.[ruder?] nach G.[nadenfeld]. Hat nichts vom Sonntag, auch schon vorigen Sonntag nicht. Etliche Gesichter, die wir vorigen Sonntag noch sahen, sahen wir heute nicht, sind weg von hier. Großpapas Nachbarhaus ist leer, ist ein anderer hineingegangen. Für viele

[xxi] Kein bestimmter Tag im Brief angegeben.

Letter

1930 10 0?[127]

We would have had a better Winter than the previous one

Dear Jacob: Since we have received two letters this week, I shall try to answer them. Sent away a letter yesterday. Received a letter yesterday together with the one for H.[128] It is Sunday today, But P.[apa] is not at home. Drove with B. to G.[129] Has nothing [no benefit] from our Sunday. Also already got nothing out of last Sunday. A few faces that we saw last Sunday, weren't to be seen today. Have left from here. Grandfather's[130] neighbour's house is empty. Someone else[131] has moved

[127] No day on the letter, but probably after October 13. [See letter 1930 10 13 re: cow & pig]

[128] Heinrich, HP 1911.

[129] G. is usually Gnadenfeld. The "B." might be brother, meaning ASR 1889's brother, David or Abram. JP 1910 in his *Memoirs* says the B stands for a Mr. Braun. In any case, a trip to Gnadenfeld, a main centre for interrogation of local officials, implies another villager was under scrutiny. Because the family still had a horse, PJ 1883 was saddled with the sad duty of transporting a friend and neighbour to his interrogation.

[130] Maybe Jacob Johann Rahn 1840-1920, whose house was in Sparrau and was now inhabited by son Johann J. Rahn 1879, but usually Grandpa refers to BB Janz 1852.

[131] Probably a communist official or party member. My aunts suggested that it was probably a Peter Guli[g]?, a member of the family that had been the village shepherds in earlier years.

eine schwere Woche gewesen. Wer weiß, was uns bevorsteht, der Boden wird leer gekehrt und das genügt nicht. Wenn es so weg geht, dann gehen wir dem Hunger entgegen. Es ist ein großer Druck unter der Menschheit. Ach wie leid tun uns die armen Kinder, wenn wir die beschauen und es scheint, ob kein Erbarmen ist. Wünschen uns so, wenn wir doch die Kuh halten könnten. Das Schweinchen haben wir geschlachtet. Es war uns wirklich schade, da war großes Gedeihen drin, aber nur füttern und würden es auch nehmen. Es hatte schon ein halbes Merka Schmalz. Es hatte Wurst. Es war schön. Wir hätten es besser gehabt als vorigen Winter. Aber jetzt liegt es dunkel vor uns. Möchte der liebe Gott sich unser erbarmen, sonst sind wir alle verloren. Wir machten drin rein, jetzt haben wir aufgehört. Erst mal sehen was die Zeit bringen wird. Haben gewaschen. Jetzt wollen wir flicken

in. For many it has been a very difficult week. Who knows what lies before us. The storage loft has been swept bare[132] but that is not sufficient.[133] If matters continue on in this fashion, then we are marching toward famine. This is a huge burden among the populace. Alas, how sad we are for the children, when we observe them and it appears that there is no salvation.[134] We wish so much to be allowed to keep the cow. We have slaughtered the pig. It was really sad, for there was great nourishment therein. But only to feed it and they would take it too. It had a half a tub (merka) of lard. It contained sausage. It was nice. We would have had a better Winter than the previous one. But now the future lies darkly before us. May the dear God have mercy on us; otherwise we are all lost. We had started to do house cleaning, now we have stopped. First, will see what time will bring us. Have done the wash! Now want to

[132] The loft was where the Mennonites stored their harvest and food stuffs for the coming winter.

[133] [In the minds of the authorities.] In the Spring most village farmers were aware that they would be expected to deliver substantial amounts of wheat grown on their private fields to the state. As a buttress against hunger and famine PJ 1883 and ASR 1889 had purchased a pig at some expense and raised it along with ducks and chickens in the hope that at least there would be meat available to them. But now that the local communists had confiscated their grain and their arbitrarily established production targets had not been reached, the entire population, even the poorer villagers, had to be squeezed for what more foodstuff might be available.

[134] [for them from the approaching famine.]

und nähen. Habe bisschen Ware bekommen. Heute hatten wir Entenbraten, dann war Großpapas Bruder hier zu Gast. Er scheltet, dass nicht geschrieben wird. Er sagt, sie sollen schreiben, dann wird er auch. Er will nach W.[aldheim] zur Sch.[wester] fahren, aber wegen der Unruhe halber bleibt er noch zurück. Franz Dirksen Geschw.[ister] sind eine Nacht verschwunden. Jetzt schon über eine Woche. Haben aber noch alles heimlich verkauft. Ihr werdet da schon vielleicht Nachricht haben, wenn Ihr diesen Brief erhalten werdet. Sie haben hier auch einen Brief erhalten als sie schon weg waren. Änni sagt immer, "Tante Anna durchgebrannt", denn ihr Haus ist vernommen. Änni ist wieder dick und gesund, macht viel Freude. Sie saß heute aufs Fenster und lass auf dem Konvertwerk von Dir....[148] Jetzt sitzt sie und malt am Tisch, auch die anderen beiden. Mika ist ein unruhiger Geist, sie bleibt Niemand was schuldig. Sie kann auch schon in der Fibel lesen. Bin mit den 3 Kleinen allein zu

[xxii] Ein Satz in der Gothischen Schrift kann ich nicht lesen.

patch and sew. Received a few [dry] goods. Today we had fried duck. Grandpa's brother was here.[135] He complains, words that cannot be written down. He says you are supposed to write [him]. Then he will [respond] too. He wants to drive to W.[136] to his sister, but because of the unrest he still holds back. The Franz Dirksen family disappeared one night. Already, more than a week ago. But had sold everything secretly. Perhaps, you already have news [of this matter] by the time you receive this letter. They too received a letter when they were already gone. Anni is constantly saying, "Aunt Anna has been burnt out" for her house[137] has been occupied. Anni is robust and healthy again, is creating much happiness. Today she sat on the window ledge and read on the envelope about you. ....[138] Now she is sitting and drawing at the table, the others as well. Mika is a restless spirit, she owes no one anything.[139] She can already read out of the primer (*Fibel*). Am home alone with the three

[135] Probably Ohm Heinrich Janz, BB Janz 1852's brother.

[136] Probably Waldheim, now called Vladovka, a village approx 10 kilometres directly North of Gnadenfeld.

[137] Aunt Anna's, perhaps Aunt Anna Jacob Rahn 1884.

[138] Anni was pretending to read about her brother JP 1910 in Canada. One sentence here is not readable and is not transliterated from the Gothic.

[139] [She makes sure that she always repays in full measure].

Hause. Hier wird gegenwärtig sehr gejägert. Tante Anna war gestern hier und sagte, sie hatte auch noch nicht Antwort auf ihren Brief von Großmama. Bei unseren Nachbarn ist es jetzt voll, der alte Vater wird wohl bald heim gehen. Ist heute sehr krank gewesen. Die Schwestern sind auch beide hier, aber die sind ganz verlegen mit ihrem Schwager. Die Schwestern und seine Frau fuhren nach Heizung und er treibt im Dorf herum. Der ist faul wie noch nie. Wenn die Schwestern nicht hergekommen wären, dann wären die tot gefroren. Die Schwestern scheuen sich nicht vor der Arbeit. Wie unseren da mal. Er ist ein fauler Mensch. Die Frauen schelten ihn schon mal wegen seiner Faulheit. Du schreibst nach Bilder. Es ist traurig, wenn

small ones. At the moment there is much hunting going on around here. Aunt Anna[140] was here yesterday and said that she had not yet received any reply from Grandma[141] to her letter. At our neighbours things are coming to a head (*ist es jetzt voll*).[142] The elderly father will doubtless soon go home [die]. Was very sick to day. The sisters are both here but completely at a loss [embarrassed] with their brother-in-law. The sisters and his wife drove off to get heating materials and he spent his time idling about the village. He is lazy like no one before. If the sisters had not arrived, then they [the neighbours] would all have frozen to death. The sisters are not afraid of work, just like our [relatives] in the past. He is a lazy person. The women sometimes scold him for his laziness. You sent a request for pictures. It is sad when not even a glass

[140] Anna Rahn 1884, never married. She lived in Sparrau with her brother, Johann 1879 and her sister, Helena Rahn Dueck 1891 and their families.

[141] Probably Helena (Vogt) Penner Janz 1854 / Mrs. Benjamin B. Janz, who was now living in Coaldale Alberta.

[142] This is the Aron Derksen family that lived next door to the East of PJ 1883. Aron had married a Margarita Dreidiger from Pordenau. She had two sisters, Lydia & Tina. They frequently came to help Margarita, since Aron seems to have been rather lazy and she had to cope with his aging parents as well as run the houshold. Apparently, Tina had been betrothed to Rev. J. B. Toews, but the marrage did not take place for Toews migrated to North America, and Tina never got out of the Soviet Union and died during World War II in Chiblinsk.

kein Glas zu haben ist. Nachbars Kajta ist mit ihrem Mann auch weggefahren, nun die Zeit wird alles lehren. Ihr wisst ja alles. Möge der Herr das Schreien seiner Kinder erhören. Nun ich werde schließen, die Kinder werden unruhig, der Schlaf kommt. Alle Abende sagen wir uns, wie schön dass wir noch in unserem Heim sein dürfen. Wer weiß was unser Los sein wird. Zum Schluss sei herzlich gegrüßt von den Lieben aus der H.[eimat]. [ASR 1889][xxiii] Du darfst Deinen Namen nicht unter schreiben, damit sie nicht wissen, von wem die Briefe sind und wir werden es auch nicht tun. Wenn wir das letzte Korn werden herabgefegt haben, dann werde ich Dir schreiben. Für die Pferde, Kuh haben wir schon nicht. Für uns vielleicht noch 25 Pud Korn, aber es soll vom Boden herab. Geld giebt es nichts mehr fürs Getreide, anfang lies ja, aber jetzt auch schon nicht mehr. Werden dir alles schreiben, wie es ausfallen wird.

[xxiii] Unten, an den Seiten, oben, über dem Datum, und inzwischen sind die folgende Zeilen von PJ 1883.

[mirror?] is to be had. The neighbour's Katja together with her husband has driven away. Well time will reveal all. You know everything. May the Lord hear the cries of his children. Now I will close; the children are getting unruly, sleep time is approaching. Every night we tell ourselves how good it is that we are still allowed to remain in our own home. Who knows what our lot will be. In closing, hearty greetings from your loved ones in the homeland. [ASR 1889]

You[143] must not use your name when signing off in order that they do not know from whom the letters come, and we shall not do it either. When we will have cleared the last of the grain [from the loft] I will write to you again. Re the horse and the cow, we no longer have them. For us there remain about 25 Puds[144] of grain, but it is supposed to be brought down from the loft. There is no money to be had for the grain. At the beginning [of collectivization] it [to sell grain for money] was still allowed, but not any longer. Will write you everything, how it turns out.

[143] The following lines were added as a post script upside down across the top of the letter & in between the lines of ASR's text by PJ 1883.

[144] 1 Pud = about 16 kilos.

Karte

1930 10 20

Da Ich nicht schlafen kann

Lieber Jakob! Da Ich nicht schlafen kann, so bin ich aufgestanden um dir etliche Zeilen zu senden. Wir sind gesund, haben auch Heute noch zu essen, ob Morgen das ist eine Frage der Zukunft. Was bei uns vorgeht, da hast du keine Vorstellung von. Es wird verlangt was der Acker nicht getragen hat. Dann wird nach gesucht und bestraft und verkauft und das hört nicht auf. Hier haben sie schon so gehandelt und werden noch mehr. Da sind nun so behandelt worden: D. W., K. R., P. R., D. B., H. S., – mit denen ist aufgeräumt, und K. T., A. T., W. J. W., W. H. D., P. P., J. B. – diese sind die folgenden. Ob wir verschont bleiben weiss Gott. Es scheint als ob nun mal Alles aus ist. In diesen Tagen sagte eine Schwester zu mir: "Ich glaube bald nicht

Post card

1930 10 20

Since I cannot sleep

Dear Jacob! Since I cannot sleep, I got up to send you a few lines.[145] We are healthy, today have enough to eat, whether we will have tomorrow is a question for the future to answer. What is happening to us, of that you have absolutely no conception. What our acreage has been unable to produce is being required of us. Then one is investigated, punished, and one's goods are sold, and that never ends. The following have been treated thus. D. W., K. R., P. R., D. B., H. S. – For these everything has been cleaned out. and K. T., A. T., W. J. W., W. H. D., P. P., J. B., these are the next in line.[146] Whether we will be spared only God knows. It appears as if now everything is at an end. During these days a sister[147] said to me, "Pretty soon

[145] I happened to translate this card at 3:00 AM when I could not sleep and got up for a drink of water and then decided to do something useful, rather than to lie in bed and stare at the ceiling. This postcard fulfills the promise PJ 1883 made in the postscript to the previous letter that when all the grain had been removed from the loft he would send more details. Obviously he was unsure whether they would also be removed for failure to contribute a satisfactory amount to the collecting of the harvest.

[146] See commentary following!

[147] A standard form of address among Mennonites for fellow believers.

mehr an Gott." Es muss erfahren werden um mit zufühlen. Zagen ja, aber nicht verzagen. Psalm 31. 23. A. P. D. von S. [Sparrau] die sind verschwunden. Ob es glücken wird? J. J. D. von S. sitzt hinter ... auch sein Bruder von H.[xxiv] Es will so scheinen als ob Psalm 50.15 nicht wahr sei? Aber wir wollen nach Psalm 72. 23 handeln durch seine Gnade. Die Deinigen. [PJ 1883]

[xxiv] Hierschau?

I won't believe in God any more." It has to be experienced in order to sympathize. Dismayed, yes, but not despondent. Psalm 31, 23[148] A. P. D. from S.[parrau] have disappeared. Will they be lucky? J. J. D. from S. sits behind [bars]. Also his brother from H.[149] It sometimes seems as if Psalm 50, 15[150] is not true. But through his grace we wish to conduct ourselves according to Psalm 72, 23.[151] Your family! [PJ 1883]

[148] *I had said in my alarm, "I am driven far from your sight." But you heard my supplications, when I cried to you for help.*

[149] Maybe Hierschau, a village about 10 kms from Kontiniusfeld.

[150] *Call upon me in the day of trouble; I will deliver you and you shall glorify me.*

[151] *Nevertheless, I remain continually with You; You hold me by my right hand.*

COMMENTARY

These may only be a few lines, but they carry important information for many an offspring of the former residents of Kontiniusfeld and Sparrau. After reading this postcard, one is not surprised that PJ 1883 had difficulty in sleeping that night. This postcard records the names of the victims of the second concerted push toward dekulakization. Until the few weeks preceding this postcard, there had clearly been unpleasant incidents throughout the village population in which houses were taken over to serve as residences and offices for party officials, both local and those brought in from outside, and property was seized or abandoned by its owners, or people had been arrested and exiled or jailed. Most of this had been done to intimidate the villagers to join the Collective or the party. The initials recorded in this card are those of owners of *Groswirtschaften* [large farms] or a granary, or a store or some other type of business who had had their households appraised earlier as recounted in the letter 1930 09 15. Since these people were the largest property owners or employers, they would be the first to be deemed kulaks. By October of 1930 it seems that the district and village soviets felt they could now move against its most powerful members and confiscate their harvests and possessions to move closer to the grain procurement quotas demanded of them by the Soviet heirarchy. Throughout the Mennonite colonies in September and October 1930 many families had been evicted from their homes and driven out onto the steppes where they were forced to dig undeground earthen homes or sod huts, called resettlement.[152] The guesses as to the identity of those indicated by the initials in this card are based on the names indicated on two village maps, one published by Schroeder in the first edition of the *Mennonite Historical Atlas* and the other published by my Aunt, Anni (Rahn) Wiens 1927 in her autobiography, *Gepriesen sei die Gnade Gottes*, 1991. These people were probably condemned as category I kulaks. The initials may stand for the following:[153]

[152] DB 8/10/1930, p. 3; MR 11/10/1930, p. 7.

[153] Compare note 106.

From Kontiniusfeld

D. W. – David Wiens

K. R. – ??? It is probably a Reimer, since the initial K is not found among the Rahns. Perhaps Kornelius, but not apparent on the maps.

P. R. – Peter Reimer

D. B. – David Born

H. S. – Heinrich Siebert, or perhaps Heinrich Spenst, but the latter was not a large farm operator.

K. T. – Kornelius Toews, his property became a club later.

A. T. – Abram Thiessen

W. J. W. – perhaps Johann Wiens or Johann Willms or their fathers whose names might have begun with W, since these names come from the map of 1940.

W. H. D. – perhaps Heinrich Duerksen, or more likely his father whose name probably began with W, for this name too is found only on the later map.

P. P. – Peter Penner

J. B. – Johann Bergen

From Sparrau

A. P. D. – Abram Peter Dueck. Both names appear separately on the maps for the same property. The property became a clubhouse later.

J. J. D. – Probably Johann or Jakob Dueck whose name appears on the early map, but has been replaced by Peter Dueck on the later map.

D. – a Dueck from Hierschau???

Brief

1930 11 13

P.[apa] ist immer in Anspruch

Will versuchen in kurzer Zeit etliche Zeilen zu schreiben. Erhielten Montag einen Brief von Dir vom 20. Oktober. Wenn Du diesen Brief erhalten wirst, dann musst Du schon einen Brief und Karte erhalten haben, denn wir schickten nicht längst einen Brief und eine Karte ab. Nun ehe ich weiterschreibe wünschen wir Dich die beste Gesundheit und das beste Wohlergehen. Gesund sind wir jetzt alle, aber vom Wohlergehen ist nicht zu sprechen. Wir haben, dem Herrn sei Dank, noch nicht so zu klagen wie schon viele bis heute. Aber was Morgen geschehen kann, wissen wir nicht. Der Boden ist leergefahren. Wir haben über 30 Pud Samen hinausgefahren und zum Pressen nichts gehalten. Es hat uns viel gekostet, aber wir entsagen schon viel, wenn wir nur Brot für den Winter über haben könnten und uns die Kuh und Kartoffeln halten könnten. Die Zukunft ist sehr dunkel. In der Natur ist es schöner. Hat

Letter

1930 11 13

P.[apa] is always under indictment

Will try in a short time to write a few lines. Received a letter dated 20 October from you on Monday. When you receive this letter, then you should already have received a letter and a card. For we sent off a letter and a card not long ago.[154] Now before I continue writing, we wish you the best of health and well-being. We are all healthy, but well-being is not to be spoken about. We have, thank the Lord, not as much to complain about as many already have by now. But what will occur tomorrow we do not know. The storage loft has been hauled empty. We have hauled out over 30 pud[155] of seed and kept nothing to press. It has cost us a lot, but we will sacrifice much as long as we might have bread for the winter and are allowed to keep the cow[156] and the potatoes. The future appears very dark. In nature it is more beautiful. It rained, now

[154] Among JP 1910's letters the postcard referred to may be the one of October 20th, the same date as the one JP 1910 wrote to them. However no letter prior to this one but after the postcard has survived. Either it never arrived, or it has been lost.

[155] Approx. 500 Kgms. or 1200 lbs. [of sunflower seeds].

[156] It seems that PJ 1883 was somewhat premature in letter 1930 10 0? where he wrote that they no longer had the cow. Now, however, they are still allowed to keep the cow.

geregnet, jetzt ist Sonnenschein, die Felder grün, aber in den Häusern ....[xxv] Es sind mehrere verschwunden, wissen nicht wohin. Hans ist Knecht bei B. D.[xxvi] sein Nachtbar. Vorigen Sonnabend war O. M. hier und sein S. J.[xxvii] Dann zweifelten sie schon, ob Du auch gekommen seist (nämlich Än.[ni]) Wenn wir gesagt hätten, dass Du das seist, dann mein ich, hätte sie das geglaubt. Wenn die anderen ihr zanken, dann sagt sie immer, du dann nicht Himmel kommen. Sie hält mit M.[ika] sehr zusammen. M. sorgt auch sehr für sie, wenn sie sich auch oft zanken. M. ist sehr kräglich. Nun ich muss abbrechen mit Schreiben. Ich weiß nicht, was ich noch schreiben soll. Zu schreiben weiß man viel, aber dann werden die Briefe nicht gehen, denn das Herz ist voll. Aber es hilft auch nicht viel, dadurch wird unsere Lage nicht besser. Haben jetzt geflickt und L.[ena] nähte sich eine Bluse. Ich Unterhosen, jetzt will ich für P. noch Hosen nähen und für die Mädchen und Än. ein Kleid. Aber man muss sich alle Lust zusammennehmen. Muss schließen, es ist Abend. Ohm H. kommt hier oft durch. Lebt noch so wie immer. Er will immer zur

[xxv] Ein paar worte sind durchgestrichen.

[xxvi] [Bruder David] David Schartner 1891

[xxvii] [Onkel Martin] Martin Jacob Rahn 1888, und [Sohn Jacob oder Johann??].

there is sunshine, the fields are green, but in the houses.........[157] Many have disappeared, do not know to where. Hans is servant at B. D.'s[158] neighbour. The previous Saturday O. M. and his S. J.[159] were here. Then they[160] questioned whether you had also come (namely An.[ni]). Had we said that you were there,[161] I think she would have believed it. If the others quarrel with her, she always says, "Then you will not go to Heaven." She regularly hangs out with M.[ika]. M. also looks after her very concernedly, even though they also often quarrel. M. is very teasing! Now I must break off with writing. I do not know what else I should write. There is much that I know that one might write, for the heart is overwhelmed, but then the letters will not reach their destination. Moreover, it[162] does not help much. Our situation will not be improved by it. Just finished patching and L.[ena] is sewing a blouse for herself. I am sewing underpants. Now I want to make a pair of pants for P.[apa] and for the girls, and a dress for Anni. But it requires that one summon all one's enthusiasm. Must end, it is evening. Ohm Hein often comes through here. Still lives in the usual fashion. He always

[157] Some words have been crossed out.

[158] Hans [JohP 1913]. B. D. is Brother David Schartner 1891.

[159] O. M. is perhaps Onkel [uncle] Martin [Martin Jacob Rahn 1888]. S. J. [son, Jacob?]

[160] The younger children, maybe Mika and Anni.

[161] One of the visitors.

[162] [to write down all the grievances]

Schw. fahren, aber es ist noch zu unruhig. Die müssen auch weg. Auch Frau J. Jansens Halbschwester muss mit ihren Kindern das Heim verlassen und so noch mehrere. P. ist immer in Anspruch. Ist gegenwärtig nach G. gefahren. H. besorgt im Stall. Die Kuh gibt noch schön Milch. Nehmen auch noch alle Tage paar Eier aus. Hätten ganz gut gelebt, wenn es nicht so gekommen wäre. Denk Dir mal 30 Pud Samen, wie viel Öl hätte das gegeben. Es gibt manches durchzukämpfen, wenn man daran denkt das Brot zu nehmen, aber wir können nichts tun, als es nur unserem Herrn zu sagen und was der für uns beschlossen hat, ist uns unklar. Ein Pud ist schon weg und das andere kommt auch weg, die sind zusammen gestellt. Man freut sich, wenn es erst Abend ist daß man über alles schlafen gehen kann. Ei, wie sieht es da? Ist da was Nachricht? Haben Dirksens schon

wants to travel to his sister, but it is still too unsettled. They have to go away as well. Mrs. J. Jansen's half-sister with her children has to leave her house, and as do several others. P.[apa] is still under indictment (*in Anspruch*). Has presently driven to G.[163] H.[164] is looking after the barn. The cow is producing milk well. Also collect a couple of eggs every day. Would have been able to live quite well, if matters had not turned out as they have. Think about it, 30 pud of seeds. How much oil that would have produced. There is much to struggle with [internally] when one considers the taking of bread. But we can do nothing about it except tell our Lord and what he has decided concerning us is unclear to us. One Pud is already gone[165] and the other will also soon be put away, they have been put together. One rejoices when it is evening and one is able to go to sleep. And how are things there? Is there any news? Have D.[avid] Dirksens received news about

[163] [Gnadenfeld]. Probably for some kind of interrogation. As we have already seen, Gnadenfeld seems to be where most of the interrogation of people from Kontiniusfeld took place. He had previously been visited by Communist officials. See 1930 04 01 and 1930 10 ?? Probably for some kind of interrogation. As we have already seen, Gnadenfeld seems to be where most of the interrogation of people from Kontiniusfeld took place.

[164] Probably Hans, JohP 1913.

[165] Probably hidden. This sentence is a bit garbled, perhaps on purpose, since she seems to be implying that they are hiding foodstuff.

Nachricht von ihren Geschwistern, hier ist nichts zu hören. Schreib uns doch, wenn ihr was hören werdet von ihnen. Justel fragt, ob ich was von ihr geschrieben habe. Die hat jetzt Ferien. Morgen fängt wieder Schule an. Sie wollen immer herumtreiben. Wir würden uns abnehmen lassen, aber ist kein Glas zu haben. Den 14. November. Will noch paar Zeilen zu schreiben. Habe eben gemolken. H. besorgt. Lena heitzt Ofen. Ju. macht sich fertig zur Schule. Heinrich ist auf seinem Posten. [ASR 1889]

P. muss zu Na. Der hat nicht Zeit zum schreiben. Ich wollte, er sollte den Brief beenden. Der ist dem Gerichte gestern übergeben, weil wir nicht ausgefüllt haben. Es haben noch viele nicht ausgefüllt. Die kaufen und fahren. Wenn wir auch die Kuh verkaufen, dann reicht es noch nicht zu zum Kaufen. Beim Bruder David schräg über die Straße wird alles weggenommen.

their siblings. Here one hears nothing. Do write if you hear anything from them. Justel asks if I have written anything about her. She is on holidays right now. Tomorrow school is starting again. They always want to romp around.[166] We would take a snapshot, but there is no lens to be had. 14 November. Want to add a few lines. Have just finished milking. H.[ans] is tending.[167] Lena is heating the oven. Justa is getting herself ready for school. Heinrich is at his post.[168] [ASR 1889] P.[apa] has to go (*Na.*)[169] He has no time to write. I wanted him to finish the letter. He was summonsed to court yesterday because we did not fulfill the required [grain] levy. There are many others who have not fulfilled what was required. They are buying and hauling (*fahren*).[170] Even if we sold the cow, we could not buy enough.[171] Diagonally across the street from

166 Meaning the little children, Justa, Mika and Anni.

167 [The cow in the barn].

168 As the next letter shows he was tending the horses during the winter months.

169 Probably *during nights* to work. He often seems to have been assigned a nightshift in later years.

170 The communists, in order to drive the non-collectivized farmers out of business, demanded large contributions of their private harvests as a form of tax. Usually this contribution was more than the individual could produce. Many people were buying grain on the black market and contributing it to the local party or collective or grain procurement commission. It seems that PJ 1883 had not provided sufficient of the harvest to satisfy the communist authorities.

171 On the black market to satisfy the demands of the levy.

Es wird sogar Mehl genommen. Was sie mit uns machen werden, wissen wir nicht. Nachgesucht haben sie bei uns noch nicht. Es wird alle Tage gesucht. Dann wird auf Stellen versteckt gefunden und so viel schlimmer ist es. Wer weiß, was uns die Weihnacht bringen wird. Sehr dunkel ist die Zukunft. Ueber uns noch sitzt einer im Regiment und schaut zu, wenn der wird Halt gebieten, dann wird es genug sein. Aber das Beten tut not, das wir erhalten bleiben. Deine Lieben in der Ferne. [ASR1889] Wenn ich DD adressiere, dann tu ich es mit Absicht. Stelle deinen Namen nicht mehr unter.[xxix]

[xxix] Diese Zeile war oben ueber dem Datum hinauf geschrieben in einer anderen Handschrift.

brother David's[172] place, everything is being taken away. Even the flour is being taken away. What they will do with us, we do not know. They haven't searched our place yet. Every day they are making house searches. Then when something hidden is found, it is all the worse for the inhabitants. Who knows what Christmas will bring us? The future is very dark. Over us there still sits One who watches. When he shall call halt, then all will come to a halt (*dann wird es genug sein*). But praying is very necessary, that we be preserved. Your loved ones in the faraway distance. [ASR1889]

If I am addressing DD, I do it with purpose. Do not sign your letters anymore either![173]

[172] David Schartner 1891 [at the neighbour's place, probably F. Derksen or H. Kasper].

[173] Both lines are written across the top of the first page of the letter upside down after the letter had already been completed. It looks like a different handwriting, probably PJ 1883.

Brief

1930 12 02

Von uns wird immer mehr verlangt

Will mal wieder versuchen, etliche Zeilen zu schreiben. Haben 14 Tage nicht geschrieben, denn es kommt uns so vor, als wenn die Briefe nicht alle übergehen. Da muss eine Karte und ein Brief kurz hintereinander hin kommen. Wir haben von Dir auch schon 3 Wochen keine Nachricht. Jetzt erhielten wir Sonnabend einen und heute einen von 9. Nov. Auch P. Penners erhielten einen von Liese B., wo das Bild drin war, wo sie im Wagen sitzt. Die Post gab die bei uns ab. Unsere Kleinen schauten, ob bei uns auch ein Bild würde hinausfallen. Änni sagt, "Jasch, fehl Apel." Die würden dich gerne besuchen bei den Äpfeln. Toben gerade sehr herum. Lena schilt schon. Hans ist im Kaninchenhandel hineingeraten. Hat sich zwei gekauft und noch gekauft und dann teurer verkauft. Wir sagen schon, er soll das sein lassen, die Jungens werden ihn anführen. Zu tun hat er wenig, nur eine Kuh im Stall. Heinrich kommt nur zum Essen nach Hause. Sie sind zu 4, die in

Letter

1930 12 02

Constantly more is being demanded

Will[174] try once again to write a few lines. Haven't written for 14 days, because it seems to us that not all the letters are being sent across. A postcard and a letter should have arrived closely one upon the other. We also had received no news from you for 3 weeks. But now we received one Saturday, and today one from the 9th Nov. P. Penners also received one from Liese B., in which there was a picture, the one of her sitting in the wagon. [When] the postman delivered the mail to us, our little ones were looking to see whether a picture would also fall out of our mail. Anni said, "Jake, lots of apples!" They would gladly visit you among the apples. Are energetically charging about at the moment. Lena is already scolding. Hans [Joh P1913] has gotten involved in dealing rabbits. Bought himself two and then more and then sold them for a higher price. We keep saying that he should let it be, that the young people will report him.[175] He has little to do, only one cow in the barn. Heinrich [HP 1911] only comes home to eat. There are four of

[174] It is interesting to note that the last few letters have not contained a salutation. It seems that PJ 1883 was suddenly frigthened to reveal any names that might be picked up by the censors.

[175] [for engaging in private business, black market].

einem Stall Pferde besorgen. Bei B. David[xxx] vor der Türe, sie wohnen da in der kleinen Stube. Du fragst nach Großpapas Nachbaren. Die sind weg von hier, die sind gegenwärtig da, wo Großpapas Schwester, Tante Franzen wohnt. Aber es geht ihnen nicht gut. Ist auch kein Wunder. So geht es manchen. Was uns die Zukunft bringen wird, wissen wir nicht. Schauen sehr dunkel hinein. Du fragst, wie viel wir haben zahlen müssen, das weiß ich nicht, aber eins weiß ich, dass der Boden leer ist. Von uns wird immer mehr verlangt.

[xxx] Schartner 1891

them who tend the horses in a single barn, before the doors[176] of B[rother] David. They [the four] are currently living there in the small room (*kleine Stube*).[177] You enquire about Grandfather's[178] neighbours. They have left here; presently they are there where Grandfather's sister, Mrs. Franzen lives. But they are not enjoying it. Small wonder! That is the experience of many! What the future holds for us we do not know. But peer very darkly into it. You ask how much we had to pay.[179] That I do not know, but one thing I do know is that the loft[180] is empty. From us constantly more is being demanded. We are

[176] Probably means that the horses have been collected and installed in ASR 1889's brother David's barn [David Schartner 1891] and the caretakers are living in the house along with the family.
[177] A technical term as well as descriptive in the sense that most Mennonite houses followed a similar pattern, and the *kleine Stube* often served as a guest room. So now the communists have installed their workers here.
[178] This is BB Janz 1852, who was still alive in Coaldale, AB, but who had lived in Kontiniusfeld. The reference is probably to the Wiens family who owned a store and lived next door to the former Janz residence. The sister is probably Anna Janz Franz b.1857.
[179] Presumably an extra tax for not having submitted the full quota of harvest grain for which PJ 1883 had been indicted at Gnadenfeld in the previous month. See 1930 11 13. The extra taxes as punishment for failing to provide sufficent quota was standard communist practice. DB 8/10/1930, p. 3.
[180] [above the house and barn where the foodstuffs for the winter were usually stored].

Wir sollen noch über hundert Pfund stellen und so auch viele. Unser Nachbar muss noch viel mehr, aber der macht sich nicht viel daraus. Der verfault beinah. Jetzt sind ihre Schwestern hier, jetzt müssen die alles tun. Die schelten sehr, sehr über ihn. Die müssen auch für Heizung sorgen und er treibt sich im Dorf herum. Die Schwestern kommen bei uns oft. Die Eltern sind beide tot. Das hatte ich schon geschrieben. Es sind liebe Schwestern. Unsere Kinder haben sie sehr lieb gewonnen. Tina wollte sich die Kinder übernehmen zu W.[eihnachten] aber es ist verhindert. Wer weiß wie wir Weihnachten werden feiern. Wenn wir bloß alle könnten zu Hause sein, dann würden wir schon sehr dankbar sein, denn wer weiß, was sie mit uns noch machen werden. Derksens erhielten heute Nachricht von Katja. Die sitzen alle hinterm Schloß, die sind alle erst genommen. Die haben viel durch gemacht, dort sind Viele untergegangen. Andere Leuten haben Morgen

supposed to steal another hundred pounds[181] as are many others. Our neighbour is supposed to provide even more, but it doesn't seem to bother him. He is almost rotting away.[182] Now the sisters are here; they have to do everything. They complain bitterly, bitterly about him. They even have to provide the heating while he loiters about the village. The sisters often come to our house. The parents are both dead. I wrote that to you before. They are sweet sisters. Our children love them very much. Tina wanted to look after the children during C.[hristmas] but it was refused.[183] Who knows how we will celebrate Christmas? If only we could all be at home, then we would be very thankful. For who knows what they will still do with us! Dirksens received word from Katja today. They are all locked up. They were all arrested first. They have endured much, many have succumbed. Other people have a

[181] By stealing she means, to rob themselves or others of grain, produce, meat,etc. As some of the other letters show there was considerable theft beginning to take place.

[182] [is so lazy]. This is probably an Aron Dirksen.

[183] The anti-religious campaign really developed after the harvest, with the renewed arresting of preachers who had been spared at the last minute in the 1 April 1930 series of arrests. MR 11/06/1930, p. 7; DB 7/05/1930, p. 4; 3/12/1930, p. 3. Now, as in the previous December, the communists were trying to ensure that no Christian festivals should be celebrated.

Hochzeit. Johann Rahns Anna mit Heinrich Tießen, Sparrau. Wünsche Dir fröhliche Weihnachten. Du schreibts, wenn wir nicht Papier haben, wirst Du dazu schicken. Das hilft nicht. Es wird Alles genommen. [ASR1889]

wedding tomorrow; Johann Rahn's[184] Anna with Heinrich Thiessen, Sparrau. Wish you a merry Christmas. You write that if we do not have paper, you will send us some. That is of no help. Everything is taken from us. [ASR1889]

[184] This Johann Rahn 1879 was a brother of PJ 1883. He was shot in 1938.

Brief

1930 12 15

Wie wir die Feiertage durchleben werden, wissen wir nicht

Wünschen Dir zuvor die beste Gesundheit und das beste Wohlergehen von Leib und Seele.Wir sind dem Herrn sei Dank gegenwärtig alle gesund. Haben vorige Woche gewaschen. Sind sehr froh dazu, denn jetzt ist vollständig der Winter eingekehrt. Gestern war großes Sturmwetter, so dass der Sturm viel Pfannen vom Dach warf. Heute ist er schon nicht so groß. P. (*papa*) ist im Soviet. Muss da wirken. Musste vorige Woche einen Tag nach Halbstadt, wussten gar nicht warum. Aber es war auch wieder gut ausgefallen. Denn in dieser Zeit ist man allen Gefahren ausgesetzt. Es war wegen Geldgeschichten und nicht wir waren gemeint, sondern P. [*papas*] Bruder H.[xxxi] Justel ist in der Schule. Heinrich ist auf seinem Posten, nahm gestern schon auf paar Mahlzeiten

[xxxi] Heinrich Rahn 1872, der elteste Sohn von Jacob Johann Rahn 1840 von der ersten Ehe mit Maria Vogt.

Letter

1930 12 15

How we will celebrate the holidays, we do not know

To begin we wish you the best of health and well-being in both body and soul. The Lord be praised, we are presently all well. Washed in the week before last. Am very happy about that for winter has now fully settled in. Yesterday there was a huge storm so that the gale ripped many tiles from the roof. Today the storm is not as great. P. [papa] is in the Soviet.[185] Has to work there. Had to go to Halbstadt one day last week, had no idea why. But it again turned out all right. For in these times one expects almost every sort of danger. It was something about money matters, and it was not us that they sought, but P.[apa's] brother H.[186] Justel is in school, Heinrich [HP 1911] is at his post,[187] yesterday took

[185] [the collective]. According to Aunt Justa, JusP 1922, *Mein Wegbegleiter Jesus Christus,* p. 13, PJ 1883 did not join the collective until Spring 1933. However, in previous letters it is apparent that Aganetha was working in the Artel regularly and HP 1911 had been put to work in the barns of the collective. Now it appears that PJ 1883 had also been put to work in the office, according to JP 1910's memoirs.

[186] Heinrich Jakob Rahn 1872-1932, PJ 1883's oldest half-brother born to Jacob Johann Rahn 1840 and Maria Vogt Rahn [died about 1877]. It must have been very sad to see oneself saved at the cost of one's brother.

[187] Presumably in the collective horse barn. See letter of 1930 12 02.

Essen mit, da er nicht immer gehen durfte. Mika und Änni wollen immer Bücher haben zum Bilder beschauen. Hans hat paar Mal geholfen beim Stroh, das wird alles zusammen gefahren. Wie wir die Feiertage durchleben werden, wissen wir nicht. Geschenke sind keine zu haben. Aber zum Backen haben wir noch bekommen. Konfekt kochen die Leute sich selbst. Die sind noch zu haben, aber sehr teuer. Ihr werdet die Feiertage anders verleben wie wir, das ist sicher. Wir wollen in den letzten Tagen backen, wenn alles ruhig bleibt. Unsere Nachbarn haben schon zwei mal Nachricht von Katja. Wenn Du den vorigen Brief erhalten hast, dann wirst du wissen. Haben D. Dirksens mal Nachricht erhalten von ihren Geschwistern? Katja und Liese Klie.[wers'] Sch.[wester] Mariechen, die haben einen Posten bekommen auch auf Lohn. Frau D. Dirksens Sch.[wester] Anna besorgt den ihre Kinder und von den Anderen

food for a couple of meals, for he isn't always allowed to leave. Mika and Anni always want books in order look at pictures. Hans [JohP 1913] has helped a few times with the straw, it is all being hauled together into one big pile.[188] How we will celebrate the holidays, we do not know. Gifts are not there to be obtained. But for baking we have indeed received something. Candies are being cooked up by the people themselves. They can also be bought but are extremely expensive. You will celebrate the holidays differently than we, that is for certain. We intend to bake in the last days[189] if everything remains calm. Our neighbours have already received news from Katya twice. If you received the previous letter, then you will know.[190] Have D. Dirksens ever received word of their relatives?[191] Katja and Liese Klie.[wers'][192] Sister Marie(chen) have obtained a position for wages. Mrs. D. Dirksen's sister Anna is looking after their children, but about the others they know

[188] Even the straw is being collectivized and JohP 1913 has also been put to work in the collective.

[189] [before Christmas].

[190] [what we are talking about, 1930 10 20].

[191] According to JP 1910's memoirs, these relatives were Abram and Anna Dueck.

[192] The intent of this sentence is confused by the writer's use of abbreviations. However the first two names are, I think intended as possessives of the following word "sister." It is the sister who along with her husband have [note the plural, another stylistic feature of this correspondence] obtained some sort of salaried post.

wissen sie nicht. Von Denen lässt sich nichts hören. Wer weiß wie es dem gehen mag. Von Dir haben wir schon zwei Wochen keinen Brief erhalten. Denken, dass der Postbote Heute einen wird haben. So weiß ich nicht was Neues. Schreib doch gleich, wenn Du den Brief vor Diesem erhalten hast, das war ein langer Brief. [ASR1889]

nothing. From them nothing has been heard. Who knows how they are faring! From you we have now received no letter for two weeks. Expect that today's mail delivery will have one. Otherwise I know nothing new. Write immediately if you received the letter before this one; that was a long letter. [ASR1889]

Karte

1930 12 22

Möchte dieses Auseinanderfahren uns vereinigen

Lieber Sohn: Will dir ein Lebenszeichen senden, das ich noch da bin. Wünsche dir das der Herr bald sein Wort einlösen möchte, was Apostelgesch. 16 Vers 13 steht. Das gab mir der Herr in jenen schweren Tagen, du wirst ja wissen. Wir sind schön gesund und wünschen dir dasselbe von Herzen. Würde dich gerne einen Brief schreiben aber da nicht Konwerte zu haben sind, und unsere alle sind, so muss ich erst eins kleben. So ist noch immer dasselbe, was die Liebe nicht hat – Ohm [Heinrich] war heute wieder hier, schlief sich aus, ist auch noch immer derselbe. Haben jetzt schon 4 Wochen keine Nachricht von dir erhalten. Ich weiß

Post card

1930 12 22, Received 14 January 1931 in Coaldale, Alberta

May this separation unite us even more strongly

Dear Son,[193] I want to send you proof that I am still in the land of the living, still here. Hope for you that the Lord will soon put into effect what is written in Acts 16:31.[194] That verse was the gift the Lord gave to me in those difficult days; you will understand.[195] We are pleasantly healthy and heartily wish the same for you. Would gladly write you a letter, but since there are no envelopes to be had, and ours have all been used up, I will have to paste one together first.[196] As usual, the same thing when there is no love...[197] Ohm[198] was here today, caught up on his sleep, is still always the same. Haven't received any news from you in four weeks. Actually, I

[193] Even in this moment when JP 1883 is trying to send warm Christmas greetings to his son, he again refuses to name JP 1910, probably because he doesn't want the censors to obtain it.

[194] *They said, "Believe on the Lord Jesus Christ and thus you will be saved, you and your house."*

[195] [what I'm referring to]. See previous letters, 1930 04 01, 1930 10 0?, and 1930 11 13.

[196] [hence, the postcard]. This is the last time that a postcard was sent. All later mail was by letter in a sealed envelope.

[197] A cryptic reference to the local regime, it was underlined in the postcard.

[198] Probably OHM Hein, Heinrich Janz, the brother of BB Janz 1852.

eigentlich gar nicht was es ist. Du schreibst doch, du wolltest jede Woche schreiben. Wir haben mitunter auch jede Woche erhalten. Kl. [Kliewers] erhielten vorige Woche mehrere Briefe und auch Weihnachtskarten, Ihr müsst immer einer von dem andern schreiben, das ist für uns besser. Hatten vorige Woche sehr großen Sturm, viele Pfannen vom Gartentor sind verschlagen. Wünsche dir noch frohe Weihnachten und Neujahr. Möchte der Liebe Gott geben das dieses Auseinanderfahren möchte uns vereinigen. [PJ 1883]

don't know what is going on; you are writing, you intended to write every week. Sometimes we received [letters] every week. Kl.[eiwers] received several letters and Christmas cards the week before last. You should always write about one another.[199] That is better for us. Last week we had a very big storm, and many tiles were broken at the garden gate. I wish you a happy Christmas and New Year. May the loving God grant that this separation may unite us even more strongly. [PJ1883]

[199] The writer is suggesting that both JP 1910 and Abe Kliewer who had travelled together to Coaldale should always both write to their families about the other. Thus if one of the letters did not get through, the news of each might still reach the parents via the other family.

Brief

1930 12 28

Haben die 3 Feiertage in aller Ruhe durchleben können.

Will wiederum versuchen, etliche Zeilen zu schreiben. Habe eben aufgeräumt. Haben den letzten Entenbraten in der Röhre. Heinrich war eben zum Frühstück zu Hause. Papa, Hans und Lena sind zur Versammlung gegangen. Justa ist in der Schule. Haben die 3 Feiertage, dem Herrn sei Dank, in aller Ruhe durchleben können. Der Herr hat die Gebete erhört und hat alles Störende zurückgehalten. Die Kinder hatten auch noch einen Heiligen Abend. Sie hatten das alles unter sich getan. Hans Töws seine Braut wollte sich das übernehmen. Aber es wurde gewarnt. Dann würden sie hier weg müssen. Jetzt übernahm sich das Susa Töws. Walls Peters Schwester, was bei Töwsen ist. Die Jungen gingen zusammen holen, es ging ganz gut. Demnach welches seine Kinder alles gemacht hatten. Der Abend war ganz schön. Hatten den gar nicht erwartet. Hier kam gerade zu Weihnachten ein Laden her. Waren auch viel Konfekte zu haben. Haben

Letter

1930 12 28

Managed to spend the 3 days of holiday in complete peace.

Want to try once again to write a few lines. Have just finished cleaning up. Have the last duck roast in the stove pipe oven. Heinrich [HP 1911] was just home for breakfast. Papa, Hans and Lena have gone to the Church (*Versammlung*). Justa is in school.[200] Managed, thank the Lord, to spend the 3 days of holiday [Christmas] in complete peace and quiet. The Lord has answered prayer and kept everything disturbing away. The children even had a Christmas Eve celebration (*Heiligen Abend*). They did that all on their own. The fiancée of Hans Toews wanted to take over [responsibility for] this [program]. But there was a warning! That it would be necessary to deport her. So now Susa Toews, Peter Wall's sister, took it over. The boys went around gathering [the villagers]. It went pretty well, given the fact that the children had done everything. The evening was very nice. Had never expected that. Here for Christmas a supply [or merchant] wagon arrived. Lots of candies

[200] The day this was written was a Sunday. Hence the gathering to which PJ, Hans and Lena went was probably a church service. That Justa was in school is the result of the change in teachers. The new teacher was a communist who was affirming the anti-religious policy of the party.

auch können Kuchen und Zwieback backen. Die Kinder bekamen Körbchen mit Blumen zu Weihnachten und Justa ein Nähtasche und Mika ??[Dom.], haben schon viel damit gespielt und Änni noch eine Puppe. Die Großen mussten sich schon so begnügen. Mit der Wäre sieht es sehr traurig, ich sagte schon, wenn es so weiter geht, dann wird es sehr traurig werden mit den Kleidern. Aber der Herr muss ja sorgen, mit unseren Sorgen erreichen wir nichts. Aber fühlen uns alle Abende dankbar, dass wir noch in unserem Bette schlafen gehen dürfen. Haben gegenwärtig schönes Wetter, kein Eis an den Fenstern und draußen schneit es. Haben lange auf einen Brief gewartet. Wenn der Postbote kam, dann lachte er schon und sagte „Sie haben wieder keinen". Bis Heiligen Abend, dann erhielten wir einen und gestern den Anderen mit Deiner Fotografie. Wir fragten Änni wer das sei, dann schaute sie, schließlich sagte sie, „Ich habe solche kleinen Augen, ich kann nicht sehen." Als wir dann sagten, dann hat sie das Bild geküsst und gedrückt. Dann sagte sie, *Jasch gef mi Apel* und küsste das Bild. Ach die würden Dich bestürmen, wenn Du hereinkommen solltest. Wenn

were also to be had. Were also able to bake cookies and buns (zwieback). The children received baskets with flowers for Christmas, and Justa a sewing pouch, Mika dominoes.[201] They have already played with them a lot, and Anni even received a doll. The adults had to be content.[202] Regarding the dry goods (wäre) it is really very sad. I said that if things don't change, then it will turn out very badly with our clothes. But the Lord must provide, with our worries we accomplish nothing. But every evening we all feel very thankful that we are still allowed to go to sleep in our own beds. Have nice weather at the moment, no ice on the windows, and snowing outside. Waited for a long time for a letter. When the mail carrier was arriving, then he was already grinning and would say, "You again have none." That was until Christmas Eve, then we received one and, yesterday, the other one with the photographs. We asked Anni who it was. She looked and finally said, "I have such small eyes, I cannot see." When we told her, she kissed and hugged the picture. Then she said, "Jake, give me apples" and kissed the picture. Oh my, how they would attack you, should you suddenly enter the house.

[201] Although the word is unreadable in the Gothic original, JP 1910's parpahrase says "dominoes."
[202] Meaning, they did not receive anything.

??[*Nachbar*] Mädchen hereinkommen, dann werden die gleich bestürmt. Daraus Dirksens Geschwister und die anderen sitzen auch. Katja Dirksen schreibt, dass die bis den 25 Dezember sollen abgerichtet werden. Weiß einer wie es ausfallen wird? Nun ich werde abbrechen. Ich hab schon sehr gescholten. Die Kinder stoßen immer an den Tisch. [ASR 1889]

Da ich mal wieder nicht schlafen kann und Mama schon geschrieben hat, so will auch ich noch ein paar Zeilen zu Papier bringen. Wünsche Dir die beste Gesundheit und noch mehr Erfolg im Lernen. Freue mich, dass Du gut lernst, aber wenn wir das von anderen auch hören, wir glauben es ja, dass es so ist, aber Du musst Dich auch nicht zu sehr loben. Nun genug von dem. Haben Heilig Abend einen Brief von Dir erhalten und letzten Feiertag auch einen, wo die Fotografie drin war. Nun zeigte ich die Änni und fragte sie, wer das sei. Dann sagte sie, sie kenne die nicht. Ich sagte zu ihr, schau mal besser, dann sagte sie, sie habe zu kleine Augen. Und dann nahm Lena sie und zeigte ihr die noch einmal und sagte, das ist Jasch. Dann nahm sie die

When [local, neighbour?][203] girls enter, they are immediately under attack. On that note, Dirksen's relatives and the others are under arrest. Katja Dirksen writes that they were to be sentenced by December 25th. Who knows how that will turn out? Now I will stop. I have already scolded [them] severely. The children are always bumping against the table.[204] [ASR 1889]

Since again I cannot sleep and Mama has already written, I too want to put a few sentences on paper. I wish the best of health and even more success in your studying. Am delighted that you are learning well. However, when we also hear it from others, naturally we believe that such is the case. But you must not praise yourself too much. Well enough of that. Christmas Eve we received a letter from you, and on the last day of the holidays another that contained the photography. So I showed it to Anni and asked her who it was. She said that she didn't recognize [the person] in the pictures. Then I said to her, "Take another look." She then said that she had small eyes. Then Lena took her and showed her the pictures again and said, "That is Jake." Then she took them

[203] The word must be *neighbour* or something similar because the neighbours on both sides were Dirksens, and thus the introduction to the next sentence, *On that note*..., makes sense.

[204] This part of the letter was written by Aganetha Schartner Rahn 1889.

wieder und schaute hinauf. Dann küsste sie die und drückte sie fest an ihre Wange und dann küsste sie die noch einmal und alles aus sich selbst, ohne dass wir ihr etwas sagten, dann ?? [*siehst*] wie lieb sie ist. Sie sprechen viel von Dir. Haben kein anderes Papier. Ist auch Keins zu bekommen. Couverts haben wir noch zwei gefunden, dann sind die auch alle und ist kein Papier und auch Couverts nicht zu bekommen und Geld schicken hilft auch nichts, denn die D. werden National [*isiert*] ???. Solltest Du was erfahren, wo der Verduftete geblieben ist, so lass mich wissen, verstanden? Habe die Feiertage in aller Ruhe verleben können. Was die Zukunft bringen wird, ist nicht zu wissen. Die Hungersnot wird doch wohl sehr groß werden, denn da sind jetzt schon solche, die sehr wenig haben. Heinrich will jetzt als Traktorist lernen, ob er wird bekommen, wissen wir noch nicht. Er und Hans haben auch so diese Kleider und zu kaufen ist nichts, so dass es mir für ihn beinahe zu kalt ist, wenn er da noch viel

again and had another look at them. Then she kissed them and hugged them close to her cheek, and then kissed them again, and all of herself, without any prompting from us. So you can see how lovable she is. They [the little sisters] talk a lot about you. Have no other paper.[205] Can't buy any either! Still managed to find two more envelopes. Then they too will be at an end. So no paper and no more envelopes to be bought. Sending money also helps nothing since the D[ollars] are nationalized.[206] Should you discover where the one who vanished into thin air disappeared to, then let me know. Do you get my drift? Managed to celebrate the holidays in complete quiet. What the future holds is unknowable. The famine will certainly become very far-reaching, since there are already many who have very little. Heinrich wants to study tractor maintenance (MTS).[207] Whether he will be accepted we don't know yet. He and Hans also have such [poor] clothes and nothing can be bought, so that I think it is almost too cold for him if he has to work so

[205] This is to explain why he is writing on the back of the paper on which Aganeta had already written on the other side. The paper is of such poor quality that the ink penetrates through to the other side and makes reading the second side very difficult. PJ 1883 wrote in pencil.

[206] Meaning, become state property, are confiscated.

[207] In 1928 both JP 1910 and HP 1911 had taken a course in tractor maintenance in Gnadenfeld. Hence Henry was now applying to work in the MTS.

draußen sein muss, so dass ich ihm mein warmes Hemd werde geben müssen. Er hatte sich schon mal sehr erkältet, nun ist er wieder gesünder. Es ist einfach ein Elend in der Gesellschaft, wo er jetzt ist, spielen sie Karten und wo er dann hinein kommt, wird's nicht besser sein. David Dierksens ihre Geschwister von Sp. sind hinter Schloss und Riegel, warten auf Nachricht wie es wird. Bis heute habe ich aber nichts Näheres erfahren. Ihr Sohn sollte sich bis 25. Nov. bei Gericht entscheiden. [PJ 1883]

much outside. So I will have to give him my warm shirt. He had already caught a very bad cold, but now he is healthier again. It is quite simply pitiful, that association [probably the people with whom he tends the horses] in which he now finds himself. They play a lot of cards there, so when he comes inside, it is little better. David Derksen's relatives from Sparrau are behind lock and bolt. Are waiting for news how it turns out. But up to today, I have discovered nothing more precise. Their son was supposed to obtain [his punishment] at his trial on Nov. 25. [PJ 1883]

COMMENTARY

In August 1930 in the Chortitza Mennonite village the local communists issued a Proclamation of the Godless and distributed it widely.[208] This document and others like it attacked the preachers for "misleading the people" and led to the revived arresting of preachers in September. Many ministers who had begun to work in the Artels during the Spring and Summer in order to provide some food for themselves and their families, were now confronted with demands that they pay for what they had eaten at public expense while working. This was, of course, a means of providing a reason to arrest the religious leaders in the various communities in the following months. Moroeover as we have already seen, the authorities began to insist that people work on Sundays while promising supplies such as flour, oil and clothes. At the same time the teachers were interrogated to ascertain their military participation, their ability to teach in Ukrainian, and their religious outlook, i.e., whether they were of the godless or not . [See the interrogation document of J. J. Thiessen, *Other Documents*] Those who would not declare themselves as godless were removed from their posts, as we have seen was the case in Kontiniusfeld. The new teachers were to teach the young people to disregard the religious holidays and replace their allegiance to their faith with a commitment to Socialist festivals and youth clubs of various kinds.

Finally, this was merely a furthering of the general policy of the Soviets revived in 1929 to encourage the population not to celebrate the Christian festivals like Christmas and Easter. The intimidation of adults who might encourage the youth to celebrate Christmas Eve was obviously part of the communist program to eradicate religious observance from their society. However, since the youth in Kontiniusfeld took the lead in preparing, planning and publicizing the Christmas Eve celebration, the Communist authorities were placed in a difficult position. They were, after all, hoping to influence the youth to abandon their cultural and religious mores in favour of the new Socialism.

[208] DB 6/08/1930, p. 4; 3/09/1930, p. 3; DB 3/12/1930, p.3; DB 26/11/1930, p. 3

In Kontiniusfeld in December 1930, when it became apparent that the youth were eager to celebrate Christmas Eve, the Communist authorities must have decided to entice the village youth through the sudden and unexpected appearance of supplies of candy, sweets and presents. Clearly the sentence "*The adults had to be content*" and ASR1889's mention of the lack of household materials in the following sentence implies that the intent behind the sudden generosity on the part of the local authorities was to entice the youth. This contrasted starkly with the callous disregard of the authorities that saw large numbers of children dieing as a result of the hardships imposed on many families through their deportation to Siberia. In fact, so many children died as a result of the exiles and resettlement schemes that the Soviets began to destroy the cemeteries in the Northern settlements to make the death toll, especially among the children, less obvious and therefore less embarassing for the Soviet idealogues and propagandists.[209]

[209] DB 17/12/1930, p. 3

1931
in Kontiniusfeld

Family name abbreviations used in discussion of the letters

ASR 1889	Aganetha Schartner Rahn 1889 – 1968
PJ1883	Peter Jacob Rahn 1883 – 1942
JP1910	Jacob Peter Rahn 1910 – 1991
HP 1911	Henry Peter Rahn 1911 – 1996
JohP 1913	Johann (Hans) Peter Rahn 1913 – 1944
HelP 1914	Helena Peter (Rahn) Pauls b. 1914
JusP 1922	Justa Peter (Rahn) Prei Geddart 1922 -2004
Anni P 1927	Anna Peter (Rahn) Wiens b. 1927

1931 in Kontiniusfeld

1931 01 22
Nach Weihnachten

Lieber Jakob, Heute erhielten wir einen Brief von Dir in welchem wir sehen, dass Du noch am Leben und schön gesund bist. Er war vom 31.12.1930. Wir sind Gott sei Dank auch noch schön gesund und am Leben. Du schreibst, wenn wir schon nicht schreiben, dann sollen die Anderen wenigstens schreiben. Heinrich und Hans, die werden schon sicher nicht schreiben. Also will ich dann weiterschreiben. Wir haben hier jetzt gegenwärtig schönes Wetter. Heute ist zum Beispiel 3 Grad warm und so ist es jetzt schon eine Zeit lang gewesen. Ist ziemlich kotig. Heinrich lernt jetzt in W.[aldheim] als Traktorist. Hans besorgt die Kuh und sitzt in der Ofenecke. Lena sagt, sie wird nicht an Dich schreiben bis Du an ihr geschrieben hast. Heute ist [Dorfs] Versammlung wegen Fleischlieferung. Es sollen von unserem Dorf 13 Kühe geliefert werden. Und nicht jeder hat eine Kuh – das andere kannst Du Dir dann denken. 23.1./Die Fleischlieferung ist so ausgefallen, jeder der das betreffende Geld liefert oder richtiger gesagt, zahlt was auf der Kuh trifft zu zuzahlen hält seine Kuh. Wenn Du schreibst, musst doch mehr schreiben, sonst ist ja immer eine Seite leer. Wenn auch Lieder oder Gedichte, es würde uns sehr freuen. Und solltest Du mal etwas Neues schreiben wollen, dann schreib den Satz von

1931 01 22
After Christmas

Dear Jacob, Today we received a letter from you, in which we read that you are still alive and in good health. It was dated 1930/12/31. We too, thank God, are alive and very well. You write [request] that if we do not write, then the others should at least write. Heinrich and Hans will certainly not write. So then I will continue [to write]. At the moment we have nice weather here. For example, today it is plus 3 [C.] and it has been thus for a long time. It is quite muddy. Heinrich [HP1911] is in Waldheim learning [to be] a tractor operator. Hans [JohP1913] is looking after the cow and sitting in the corner near the stove. Lena says she will not write to you until you have written to her. Today there is an assembly concerning meat submission. Our village is supposed to contribute 13 cows. And not everyone has a cow – [so] the rest you can then figure out for yourself. 23/1: The meat delivery turned out as follows. Everyone who pays the required amount of money, or more rightly said, pays the total value of the cow keeps his cow. When you write, you should write more. Otherwise, there is always one side empty. Even if songs or poems, they would be very pleasing to us. And should you wish to write [send] us some news, write the hinten, dann meinen sie es sei Englisch, also ich werde mal schreiben. Ist für uns Hoffnung. Gnunffoh snu rüf tsi. Rückwärts geschrieben. Nun wird Mama noch weiterschreiben. Siehst Du auch, dass dies der Bogen ist, welchen Du schicktest?[i] Also will ich schließen, verbleibe Dein für Dich betender. [PJ 1883] Weil Papa schreibt, dass ich noch schreiben werde, so werde ich noch etwas schreiben. Hans trug eben Änni zu Dirks. Die kleinen gingen zu Liese spazieren. Justel hat heute Ruhetag. Änni ermahnt Heinrich oft wegen Rauchen. Sie sagt zu ihm: „Hein, wenn Du schmöckst, dann kommst Du nicht in den Himmel" oder „Hein, Schmöcken es Sünd", aber lieben tut sie ihn sehr. Gestern haben wir Tee zum Abendbrot gekocht, dann hatte sie schon mit uns gegessen. Dann kam Heinrich aus der Dienstversammlung und aß, dann trank sie noch drei Tassen Tee aus. Was übrig blieb, das räumte sie mit auf. Verkauften Arbusensamen. Das kostet 1 Rubel 60 Kopeken das Kilo, dafür holte Heinrich Zuckersand. Schade

[i] JP 1910 hat in seine Memoiren geschrieben das diese Zeile war eine Code das er mehr Papier und Couverte schicken sollte.

sentence backward, then they [the censors] will think it is English. I will demonstrate. For us there is hope. Epoh si ereht su rof – written backward. Now Mama will write further. Did you notice that this is the sheet that you sent?[1] Well, I will quit, remain the one who prays for you. [PJ 1883]

Because Papa has indicated that I will write, then I will add something. Hans has just carried Anni to Dirksens. The little ones went to visit Liese. Justa has a rest day today. Anni often admonishes Heinrich concerning his smoking. She says to him, "Hein, if you smoke, then you will not go to heaven" or "Hein, smoking is sin,"[2] but she loves him very much. Yesterday, we brewed tea for supper; she had already eaten. Then Heinrich came from the assembly for work assignments (*Dienstversammlung*) and ate. So she drank another three cups of tea. What was left over, that she helped clear away as well. Sold watermelon seeds. They cost 1 rubel 60 kopeks per kilo. Heinrich then got granulated sugar (*Zuckersand*) for it.[3] Sad that we didn't

[1] This is code for "Send us more paper", according to JP 1910's memoirs. The meaning is that the family have used up all their own paper now and must use what he has sent them.

[2] Written in Low German.

[3] [for the money earned by the sale of the seeds]. Because HP 1911 worked in the Artel caring for the collectivized horses and was also accepted into the MTS he had some opportunity to obtain foodstuffs either as wages or for money.

dass wir nicht all den Samen gehalten haben, dann hätten wir viel verkaufen können, aber wer wusste das? Hast Du die Photografien erhalten, wo Deine Mama und ihr oben seid? Die schickte Papa mal hin. Die Öl, von was ich im vorigen Brief erwähnte, haben wir nicht erhalten. Heinrich wird vielleicht bekommen, weil er in Waldheim ist. Heinrich schon darüber weg. Sollten sich damit freuhen mit dem ungewohnten Gut. Wenn wir nur unsere Kuh halten könnten, die wird vielleicht bis zum 10. Februar Milch werden. Darauf freuen wir uns schon sehr. Tante Isaak sagte, wir sollten uns mal von ihr holen, aber habe nicht Mut der ??? Hier fuhren schon mehre weg was nicht ???, sich Arbeit suchen. Mein Bruder David und seines gleichen noch mehr. Die Menschen sehen nur es geht und doch wird mit dem Strom mitgegangen. Wir können es gar nicht genug schätzen, dass wir noch zusammen sind. Wie viele müssen schon schmachten. Sind wir besser. Nun

keep all the seeds, then we would have been able to sell a lot. But who knew that? Did you receive the photographs, on which you and your mother appear?[4] Papa sent them off previously. That oil, concerning which I made mention in the previous letter, we never received. Heinrich may perhaps receive some, for he is in Waldheim.[5] Heinrich went looking for it. Well let them enjoy their unaccustomed good fortune. If only we might be allowed to keep our cow! She will perhaps freshen [calve] the 10th of February. For that we are very happy. Mrs. Isaak says that we should get [some] from her but [I] have no heart to ???

Here several who have no [rights] drove away to look for work. My brother David and his ilk even more.[6] People only see that it [the new system] works and so they go with the current. We simply cannot value it enough that we are still all together. How many already must live in reduced circumstances

[4] JP 1910 did receive it and passed a copy on to each of his children. See *Plates*.

[5] Waldheim seems to have been something of a party administration centre, particularly for the MTS [tractor service]. The writer thinks that since Henry is now nearer the source of influence and earning some rations for his training he might get some of the sunflower oil back for the family.

[6] According to JP 1910's Memoirs, this uncle, David Schartner 1891, lost his voting rights sometime during this year. His was designated a Groswirtschaft by Anni Rahn Wiens 1927 on the map in her autobiography. However in later correspondence [e.g., 1934 06 03] he and the family still seem to be present.

Gottes Gnade ist es. Schade, dass Du die Briefe nicht regelmäßig erhältst. Wir haben jetzt von Dir den 5. Brief von Heilig Abend. Nun werde zum Schluss kommen, es ist Mittageszeit. Haben Kartoffeln abgeschält. Papa hat Lesung und Joh. U. Ohm ist unser Stammgast, der kommt alle Tage. Es fängt an zu frieren. Der Kuh war ??? [ASR 1889]

(*schmachten*)! Are we better? Well it is the grace of God. Sad that you do not receive our letters regularly. We now have the 5th letter from you since Christmas Eve. Will bring this to an end. It is mid day. Have peeled potatoes. Papa has a reading (*Lesung*) and John U.[7] Ohm [Heinrich] is our resident guest. He comes every day. It begins to freeze. The cow was ???? [ASR 1889]

[7] The meaning of "reading" (*lesung*) is difficult to ascertain. This is probably a court hearing to investigate religious activity since the hearing is linked with a JU. There is a Johann Unrau on the maps. In the memoirs to another letter JP 1910 says that JU was actually Preacher Unrau.

COMMENTARY

The Ukraine had not submitted its full quota of grain from the harvest of 1931 demanded by Moscow because much of the harvest had either been brought in too late and so had rotted in the field or, because storage facilities at the stations were inadequate, had rotted or sprouted beside the railway tracks.[8] Then in order to fill the quota better the local councils sent around collectors to confiscate all the grain stores that had been set aside to feed humans and animals in the winter months as well as to provide seed for the coming spring.[9] Moreover, local authorities either because they were so busy with grain collection, or because they had surrendered all the confiscated seed grain failed to order the seeding of the winter wheat in time. Thus according to one estimate, at least one million hectares through out the Soviet Union were left unplanted.[10] Then early in 1931 new measures were put in place through out the Ukraine to fill the shortfall. New demands for taxes, wheat, chickens and cows were imposed on the individual villages. Given the Soviets' refusal to recognize the faults in the system while trying to prove its superiority to the free enterprise economy, they chose to blame residual kulak attitudes that still remained in the peasant villages. Hence, the private farmers and preachers were the ones who received the most onerous new requests.[11] At least sometimes these demands were discussed in a village assembly such as the one mentioned in the first part of this letter. Of course, since most of the harvest had already been surrendered or confiscated, the only method left to satisfy the authorities was to pay money. That of course is precisely what PJ 1883 did to keep his cow. The selling of the watermelon seeds is an example of one of the ways of obtaining the required money. Moreover the surrender of the oil that had been pressed out from the their private garden of sunflower seeds may have been another means of satisfying the authorities' demands. However, ASR 1889 saw it simply as the theft of what rightfully belonged to the family.

[8] DB 3/12/1930, p. 6.
[9] DB 31/12/1931, p. 3.
[10] DB 17/12/1931, p. 3.
[11] MR 4/04/1931, p. 6.

Brief

1931 03 05

Wenn Du diesen Brief erhältst, dann sitzen wir schon ohne Brot

Lieber Jakob, Versuche Dir paar Zeilen zu senden. Erhielten diese Tage wieder einen Brief von Dir. Wir sind Gott sei Dank noch schön gesund und wünschen Dir dasselbe von Herzen. Wir hatten hier schon paar Tage schönes Wetter, aber jetzt ist wieder Frost. In diesem Jahr haben wir viel Wind. Fünf Wochen hatten wir furchtbar großen Sturm, habt Ihr auch das Zeichen im Mond gesehen? Es soll im Mond ein Kreuz gewesen sein und um den Mond ein Regenbogen. Es haben Viele gesehen. Ich selbst habe es nicht gesehen, aber ein treuer Bruder Jordan aus M[un]tau hat es auch gesehen, deine Onkel kennen den Bruder gut. Sonst würde ich es nicht schreiben, aber nun glaube ich es auch. Hier hat es großes Aufregen erregt. Es war vom 9. auf den 10. Februar. Aber wir haben hier nachher auch noch Furchtbares erfahren. Manch einer hat auch ein Kreuz erhalten. Wieviele sind ohne Brot ? Wieviele sind ohne Obdach? Die Zahl ist groß. Ohne Obdach sind Abr. seine Eltern, A. J. T., P. Is. P., M.

Letter

1931 03 05

When you get this Letter we will be without Food

Dear Jacob, Am trying to send you a few lines. Received a letter from you during these last days. We, thank God, are all well and heartily wish the same for you. We have already had a few days of nice weather here, but now there is frost again. This year we are having a lot of wind. For five weeks we had frightfully huge storms. Have you seen the sign in the moon? There is supposed to have been a cross in the moon and a rainbow around it. Quite a number have seen it. I myself did not see it but a faithful brother, Jordan, from M[un]tau has also seen it. Your uncles[12] know him well. Otherwise I wouldn't relate this, but now I believe it as well. Here it caused a lot of excitement. It occurred [on the night] of the 9th to the 10th of February. But afterward here we experienced dreadful things. Many a one has also received a cross [to bear]. How many are without food? How many are without shelter? The number is large. Without shelter are Abr[am]'s[13] parents, A. J. P., P. Is. P., M. M. W., P. J. Duerk., D. D.

[12] B. B. Janz [1877] and J. B. Janz [1884], who had emigrated to Coaldale, AB. This phenomenon was widely reported among the Mennonite villages in the Molotschna although the report in the *Runschau* [MR 8/04/1931, p. 6] gives the date as February 2.

[13] [Kleiwer, with whom JP 1910 emigrated to Canada].

M. W., P. J. Dürk, D. D. L., K. J. T, P. G. R., K. K. R, NF. G. D. Nebst Gruß, Dein für Dich betender. Deine Schwestern singen dieses Liedchen immer zusammen, also die beiden Kleinen:

Von Land zu Land, von Haus zu Haus,
durch Sonnenhitz und Sturmgebaus,
von Schmutz befleckt, von Staub bedeckt,
so zieh'n wir ein und aus.
Ihr lieben Leute fromm und gut,
Ihr wisst es nicht wie weh es tut,
*wenn man in Not um fremdes Bro*t
muss flehen mit schwachem Mut.
Wer noch sein Vater nie vergisst,
*die treue Mutter liebt und küss*t
*Kommt Kinderlein, seht groß und klei*n
Wie arm ein Waislein ist.

So wird es in diesen Jahren manchen gehen. Uns haben sie auch das letzte Maehl genommen, so dass wir jetzt von dem vorher gebackenen Brot haben. Ich sagte immer, sie sollten nur backen, wir würden es schon essen und so teilen wir es uns jetzt zu, aber wie lange wird es reichen? Wenn Du diesen Brief erhältst, dann sitzen wir schon ohne Brot. Ich wundere mich nur, dass nicht mehr

L., K. J. T., P. G. R., K. K. R., NF. G. D., With greeting, the one who prays for you. Your sisters, namely the two smallest ones, are always singing this song together:

From land to land, from house to house,
Through heat of sun and storm's onset,
Spattered with dirt and covered with dust,
We travel in and out [here and there].
You dear people, pious and good,
*You do not know how deep it hurt*s
*When one in need for strangers' brea*d
Must beg with flagging courage.
You, who ne'er forget your fathers,
Love and kiss still your mothers,
Come Children, great and small,
See how poor an orphan is.

That is how many a one will fare this year. From us they have also taken the last of the flour, so that all we have now is the previously baked bread. I always said, they [the household] should just [keep on] baking, we would surely eat it.[14] So now we divide it. But how long will it last? When you receive this letter, we will already be without bread. I am only surprised that more people are not going mad. For we have just learned

[14] These confiscations of the last of the flour caused many people throughout the Molotschna who had heard of this occurring in other villages to try to bake everything into bread before it was seized. DB 22/04/1931, p. 4.

Menschen den Verstand verlieren? Denn wir haben wieder etwas erfahren. Mehr als anno 21 bei Abr. seinen Eltern in der großen Stube. Dein Großpapa hat eine kleine Vorstellung davon. Schreib doch deinen Namen nicht immer unter. [PJ 1883]

of something else. More people than in the year [19]21 are in Abr.'s [Kleiwer] parents' house in the big room (die grosse Stube). Your Grandfather[15] has a small inkling what that is like. Don't always add your signature underneath! [PJ 1883]

[15] B. B. Janz 1852 who had emigrated to Coaldale, AB. Obviously this is a reference to the famine and suffering at the beginning of the previous decade, the 1920s, after the intial attempt at collectivization.

COMMENTARY

This letter was written with a very heavy heart. PJ 1883's first line emphasizes that he is trying to write. Not only must he report the expulsion of many fellow villagers, but also the seizure of the last of their flour that was supposed to be their lifeline until new crops and vegetables could be grown. This was the aftermath of the earlier demand from Moscow for more grain and meat referred to in letter 1931 01 22. By setting the new procurement taxes so high that virtually no one could pay them, any and all villagers were now open to prosecution. Thus the Soviet regional and village councils had a free hand to prosecute whomever they wished.

The people referred to by these initials were mostly owners of *Grosswirtschaften*, all designated Kulaks, or sons of Kulaks, and hence the most well to do, who would be the first to suffer.

AJT – Perhaps Abram Thiessen, who owned at least two houses.

PIsP – Perhaps Peter Penner.

MMW – Perhaps a Wiens, they had had a granary at some point. Another Wiens family had owned a store, but these had already been removed earlier, see 1930 09 15 and 1930 10 20. Or a relative of Johann or Nikolai Wiebe or Johann Willms, all *Grosswirtschaften*.

PJ Duerksen – Peter Duerksen, perhaps the one who had had a milk station

DDL – Perhaps a Loewen from Sparrau

KJT – perhaps Kornelius Toews. His house became a club later.

PGR – Perhaps a Peter Reimer connected at some point with a kindergarten that might be viewed as a position of influence by the communists, also seems to have been viewed as a *Grosswirtschaft*. There is also a Gerhard Regehr in Sparrau around 1940, so this might be a son or ancestor. I do not think it was a Rahn, since the letter G does not appear among the ancestral Rahn first names at this time.

KKR – Again not a Rahn, the first names do not work, but perhaps a Rempel from Sparrau or a Reimer from Kontiniusfeld.

Nf GD – Here the initials seem to be reversed, Neufeld, G. D. The majority of the Neufelds lived in Sparrau, in particular a David Neufeld, again a son or ancestor?

In the months of March and April many of the expropriated properties including the dwellings of some of these exiled people were turned into stables for cows, horses and pigs throughout the Molotschna.[16] It is interesting to note that PJ 1883 finds refuge in a song sung by his daughters that he undoubtedly taught them. Acording to the surviving Aunts, Lena Rahn Pauls 1914 and Anni Rahn Wiens 1927, their father was a very musical person with good singing ability.

[16] DB 19/08/1931 p. 5.

Brief

1931 03 31

Brüderliche Liebe

Lieber Bruder, ich will ein paar Zeilen an Dich schreiben, weil Du an mir geschrieben hast. Wir haben noch kein Getreide in der Erde gesetzt. Ich habe für Zuckerrohr Samen abgekratzt zum Setzen. Und schreib mir auch dann gleich zurück, an mir und dann schick mir einen Dollar. Ich wünsche Dir die beste Gesundheit, hier ist das Wetter noch kalt. Habe für zwei junge Kaninchen mit eine alte Sau eingetauscht und jetzt wollen die Leute mir die abkaufen und das tu ich nicht. Der Johann nach welchem du fragst ist Abraham Franz sein Bruder, O.[hm] Heinrich will nach uns einen Brief schreiben und er kommt gar nicht fertig damit. Ich habe mir eine Taube aufgefangen und die ist fort geflogen und jetzt habe ich keine. Aber legen so [das Netz??] das sie dem nicht [sehen??]. Schick mir einen Konverk u[nd] ein Bogen Papier dass ich wieder schreiben kann an dir u[nd] schick mir ein Paar Schubaender und M.[ika] auch. Heinrich wird das übrige voll schreiben, sonst wird das Papier zurück kommen. Aber er ist auch der ??. [Johann P

Letter

1931 03 31

Brotherly love

Dear Brother: I want to write a few lines to you since you wrote to me. We have not yet put any seed grain in the earth. I have scraped off some seed for planting the sugar cane. And reply to me immediately and send me a dollar. I wish you the best of health. Here the weather is still cold. Have obtained a pair of rabbits for an old sow and now people want to buy them from me but I am not going to do that. That Johann whom you asked about is Abraham Franz's brother. Uncle Heinrich [Ohm Heinrich] wanted to write a letter after us, but he will never finish it in time. I caught a pigeon, but now it has flown far away and I have none. But am setting the [net, trap] so they [the pigeons?] don't see it.[17] Send me an envelope and a sheet of paper so that I can again reply to you a.[nd] send me a pair of shoelaces and Mika as well. Henry will fill up the rest of the paper, otherwise the letter will come back.[18] After all he is the ???[19] [JohP

[17] This sentence is not completely legible in the original letter.

[18] This is meant as a joke. The brothers are having some fun at each other's expense. Hans is teasing Henry perhaps about his studying as a "learned one"???

Rahn 1913: JohP 1913] Unser Hans denkt was hier nicht zu haben ist das kannst du ihm schicken. Er schreibt auch Keines weil er nichts zu schreiben hat. Dieses nicht mal voll geschrieben! [HP 1911]

1913] Our Hans thinks you can send him what is not available here. He writes nothing because he has nothing to say[write]. Didn't even fill this up. [HP 1911]

[19] After the signature which is in large initials, Henry draws a wiggly line under Johann's letter and adds the words that followed.

COMMENTARY

It is interesting to note that they had reached the end of March and seemed to be behind in the normal routine in their region for the planting of crops. This reflects the control now exercised by the regional or the village soviet council or the kolkhoz which through its confiscation of the horses, machinery and seed grain not only set the farming schedule, but also controlled it. Several letters in the *Bote* and the *Runschau* refer rather sarcastically to the Soviet view that the village farmers no longer knew when or how to repair their machinery, or prepare the soil for seeding,[20] or even when Winter would arrive.[21] The soviet concern to collectivize and then standardize all agriculture had led them in the previous year, at least in some villages, to declare that winter was going to set in on the 25th of October. Now obviously somewhere in the USSR that did occur. However, driven by the fear of individual initiative the Soviets' bureaucracy failed to make sufficient allowance for differences in climate in different regions resulting in many jokes about the all powerful Soviet leadership.

The references to the trade in rabbits and the catching of pigeons reinforces the notion that all the members of the family were engaged in fighting the encroaching famine. Its severity is illustrated in that during the month of April 1931 the authorities, probably the District Control Commissions, sent accredited hunters into many Mennonite villages to comb them systematically and shoot all the dogs and cats. The carcasses were then collected and delivered to the authorities to provide meat for the city populations.[22] Finally, the request for pairs of shoe laces reveals how little of their daily needs was available through purchase to these hard working villagers.

[20] MR 04/03/1931, p. 6

[21] DB 16/12/1931, p. 4.

[22] DB 8/04/1931, p. 3.

Brief

1931 04 02

Die Traktoren arbeiten nur schlecht

Guten Morgen lieber Bruder, Wünsche Dir die beste Gesundheit in der Ferne. Will versuchen, ein paar Zeilen an Dich zu richten. Es geht hier nicht sehr gut, weil es so ist wie es ist. Haben Sonntag die Traktoren von Waldheim geholt 2 John Diere und 2 Internationale. Die Traktoren arbeiten nur schlecht, sie sind wohl remontiert aber es ist nicht viel, denn es ist nur altes auf einen Tag zum Säen gemacht. Auf dem Feld haben wir noch nicht gearbeitet, denn es ist hier noch alle Tage kaltes Wetter. Aber so ist hier noch viel zu tun. Bei dem Traktorsonntag werden die Traktoristen wohl keinen haben, denn es ist sehr viel Arbeit. Hier ist noch alles beim alten. Neues gibt es hier nicht. Man hört nur Traktor und Fünfjahresplan, weiter nichts. Muß schließen, will arbeiten gehen. Grüße Heinrich Rahn

Letter

1931 04 02

The tractors work only poorly

Good Morning Dear Brother, Wish you the best of health there in the distance. Will try to direct a few lines to you. Things are not going very well here, because it is as it is. Sunday we brought the tractors from Waldheim, 2 John Deere and 2 Internationals. The tractors work only poorly, they have indeed been refitted, but not very well, since it [the refitting] has been done with old materials for the purpose of seeding for a day. But we haven't worked in the field yet, because here we still have cold weather every day. Nevertheless there is plenty to do here. As for the tractor Sunday, the operators probably won't have any [Sunday], for there is much work to do here. Here everything is done in the old manner. Nothing new is available here. One only hears of the tractor and the five-year plan, otherwise nothing. Must close, want to go to work. Greetings, Heinrich Rahn

COMMENTARY

In the letter of 1930/12/28 PJ 1883 wrote that Heinrich was applying for training as a tractor operator in the MTS. Part of the motivation was probably that he would obtain a uniform. Since clothing material was not readily available, this was a big drawing card. Moreover, Heinrich [HP 1911] and my father [JP 1910] had studied tractor maintenance in Gnadenfeld in 1928 while JP 1910 was still living in Kontiniusfeld. [He mentioned this training to me verbally, but also recorded it in a short written reminisence dated 24/05/1976.] Since Heinrich mentions that he participated in bringing the tractors to Kontiniusfeld, he obviously had received this new responsibility. Probably the tractors were originally imported via Klippenfeld/Stulnevo railway station, and were then sent on to Waldheim, from where they had been collected by the Machine Tractor Station [MTS] in Kontiniusfeld of which Heinrich was now a part. The proximity of Waldheim to the railway station may explain why it rather than Gnadenfeld [traditional regional capital] seems to have been the main regional headquarters [Eastern Molotschna] of the Machine Tractor Station [MTS].

The reference to Tractor Sunday is indicative of the sloganeering that the communists engaged in to give their new intiatives legitimacy. The casual observation that those working in the MTS will probably not have *any Sunday* [no day of rest] points to the anti-religious policy of the communist leadership. In early March another slogan, *Next Easter we will celebrate without preachers*, had been making the rounds in the Mennonite villages.[23] Moreover, a directive was sent to all teachers that school sessions were not to be interrupted during the Easter festival days and that any teachers who disobeyed the order would be summarily dismissed.[24] Within the month all teachers were expected to enrol in the *Circle of the Godless*. Many of the preachers who had worked in the Artels in the preceding Autumn and/or Winter in order to receive food rations were now ordered to return what they had been given, an obvious impossibility since most of it had been

[23] DB 4/03/1931, p. 3.
[24] MR 13/05/1931, p. 7.

eaten. Thus the authorities again had an excuse through these trumped up violations to arrest and deport the preachers. To supplement this attack on the religious leadership of the Mennonite villages, the district or village control commissions frequently held village assemblies or work planning assemblies on Sunday mornings with attendance required,[25] or as described in the present letter, a special event, *Tractor Sunday*, to attract the village population away from their church services. This Tractor Sunday was also trumpeted throughout other Mennonite villages and was obviously designed to detract from the Easter week. Even in villages where there was no local MTS, tractors were driven through to demonstrate the Soviet success story. Not surprising that Heinrich, unimpressed by the continuous reiteration of the wonderous advances of Soviet technology [the tractors after all are built by American companies], must close because he has to go to work. The truth about this attempt to replace the horses with tractors is that even in 1933 the Soviet Union had only about two-hundred thousand tractors for all the millions of acres it needed to cultivate, while the horse population had shrunk from thirty-four million in 1929 to 16.6 million in 1933.[26]

[25] DB 8/04/1931, p. 3.
[26] McClellan (1986), p. 123, 125.

Brief

1931 04 10

Leben wollen wir noch so Lange der Odem ist

Lieber Jakob, Da wir diese Woche noch keinen Brief abgeschickt haben, so will ich versuchen, ein Lebens und Liebeszeichen an Dich zu senden. Wir haben nach Weihnachten noch alle Woche geschreiben, außer einmal haben wir in 14 Tage nur einmal geschrieben und dann nimm nur mal alle deine Briefe zusammen und dann wirst Du sehen, dass Du alle Woche einen erhalten hast und wenn nicht, dann hast Du sie eben nicht erhalten. Nun ehe ich weiter schreibe, wünsche ich Dir Gottes reichen Segen und eine schöne Gesundheit in deinem neuen Dienst. Wir dürfen uns auch noch immer einer schönen Gesundheit erfreuen, dem Herrn sei Dank dafür. Heute ist ja bei uns Karfreitag, erhielten auch heute einen Brief von Dir, wo Du für die drei Kleinen die Bilder hineingelegt hattest. Wir haben noch alle Briefe von Dir erhalten, nur den nicht mit dem Haarband, die beiden Fotografien sind in unserem Besitz. Ich schrieb Dir, Du solltest uns keine Pakete schicken, weil wir nicht Geld haben zum Auslösen, aber ich befürchte, wir werden uns jetzt bald kein Brot kaufen können, weil es nicht zu haben sein wird, sonst die Preise steigen furchtbar, hier bei uns kostet das Pud Weizen schon von zwanzig bis vierzig Rubel und in der Krim

Letter

1931 04 10

Want to stay Alive as Long as there is Breath

Dear Jacob, Since we have not yet posted a letter this week, I will try to send you a token of life and love. After Christmas we have written every week except during one 14-day period when we wrote only once. If sometime you gather all your letters together, then you will see that you have received a letter every week. And if not, then you have simply not received them. Now before I continue, I wish you God's richest blessing and fine good health in your new service [post]. We also are still able to rejoice in fine good health, thank God for that. Today, of course for us, is Good Friday. Received a letter from you today in which you had put pictures for the three little ones. We have received all the letters from you save the one with the hair band, the one with both photographs is in our possession. I wrote that you should send us no packages because we did not have the money to pay the duty. But I fear that soon we will not be able to buy any bread, because it will not be available. Generally prices have risen dreadfully. For us here a Pud [about 16.4 kgms.] of wheat costs about 20 to 40 Rubels and in the Crimea up to 200 Rubels per Pud.

bis zweihundert Rubel das Pud. Es ist kein Märchen sondern die nackte Wirklichkeit und dann kommt ein Paket auslösen schon nicht so teuer und leben wollen wir da noch so lange der Odem in einem drin ist. Aber ich würde Dir raten, nur Reis zu schicken, dann werden wir schon ohne Brot leben bleiben. Solltest Du aber nicht direkt an uns schicken können, dann schicke nach Deutschland an Br. Jakob Kröker, Werningerode a/Harz, Kaiserstraße 4 mit der Bitte, er möchte es an mich senden. Das würde er schon tun. Aber nur Reis, ich meine, da kommt der Zoll das billigste. Mit der Saatzeit haben wir noch nicht begonnen. Die Partei wollte, wir sollten jetzt die Feiertage säen, aber da hat der liebe Gott ihnen dochmal wieder einen Strich durch die Rechnung gemacht und ich glaube Er wird auch Ostern dafür sorgen, dass wir die Feiertage feiern werden können. Man hat auch wirklich kein Interesse am Säen, denn was hilft alles arbeiten, wenn man nur für all die Arbeit fürs ganze Jahr hungern muss, so dass sie alles wegnehmen. Freuen uns auch sehr zu dem Papier, nur schade, dass Du aufgehört hast Kuverte zu schicken. Hier ist beides nicht zu haben. Danke schön dafür. [PJ 1883] Weil schon geschrieben ist, so will ich auch noch ein paar Zeilen schreiben. Haben Abend gegessen. Hatte von Schlichtmehl Bulka gebacken und Butter dazu. Es hat gut geschmeckt, aber es wird immer zugeteilt, wenn wir es nicht schon mussten, wenn wir

That is no fairytale, but the naked truth and then [under those circumstances] paying duty on a package does not seem so dear; and we still want, of course, to stay alive as long as there is breath in us. But I would advise you to send only rice, then we will surely survive without bread. Should you be unable to send directly to us, then send [it] to Germany, to Br. [other] Jacob Kroeker, Werningerode am Harz, Kaiserstrase 4 with the request that he should send it to me. That he will certainly do. But only rice, I think, for then the duty is the lowest. With the seeding we have not yet begun. The Party wanted that we should sow during the [Easter] celebration days, but there our gracious God has again drawn a line through their calculations, and, I think, on Easter Day He will also see to it that we will be able to celebrate too. One really has no interest in seeding, for what good is all that working, if in return for all that work one has to starve for the whole year, for they take everything away. Am very pleased to get the paper, only sorry that you have stopped sending envelopes. Here neither is to be had. Thank you very much for it.[27] [PJ 1883] Since something has already been written, I want to add a few lines. Have eaten this evening. Baked bread (*Bulke*) from plain flour (*Schlichtmehl*) [mainly starch], with butter. It has tasted very good, but it always has to be divided up. If only we didn't need to do

[27] What follows is in different handwriting.

dann nur immer hätten. Die Mädchen bitten, sie wollen auch schreiben und ich habe nicht Zeit mich mit ihnen abzugeben. Haben sich sehr zu den Bildern gefreut. Sie wollen Dir dafür immer Briefe schicken. Ich sagte, Ostern könnten sie schreiben. Jetzt sind sie wohl beim Eier färben. Du schreibst wie froh wir sein können, dass wir noch im Heim sind. Das besprechen wir auch fast alle Tage, aber fragend schaut man in die Zukunft. Ostern ist Mika ihr Geburtstag. Die Deinen. [ASR 1889]

that, if only we always had it [food]. The girls are asking to write as well, but I do not have the time to spend with them. Were very happy with the pictures. They always want to send you letters in return. I said that they could write on Easter day. Right now they are content with coloring eggs. You write how fortunate we are to still be in our home. We express that almost every day, but we peer questioningly into the future. Easter is Mika's birthday.[28] Yours. [ASR 1889]

[28] April 12 was her birthday. The date of Easter was April 5 by the Gregorian calendar, but the letter is dated April 10, Good Friday, by PJ 1883. Under the Julian Calendar Easter fell on April 12 in 1931.

COMMENTARY

This letter reflects the effect that famine can have on the decision-making process. In earlier letters, JP 1910 had been asked not to send any food parcels, since the duty that the Communist State was charging on the delivery of such parcels was so exorbitant that the receiver could not afford to redeem the food parcels. Now, however, there is a desperation in the letter's cry for help. The request to send rice is clearly an attempt to maximize what little financial resources they still have available to them. The reference to a Jakob Kroeker, emphasizes the importance of getting food delivered to them.[29]

This Jakob Kroeker was born in the Molotschna in 1872, grew up in the Krimea and was educated in Germany in an evangelical theology school under the influence of a Dr. Baedeker. He returned to teach in the Krimea, and then ran a printing press in Halbstatt. In 1908/09 he moved to Germany to Werningerode a/Harz. From there he began to travel in an evangelisation effort supported by ecumenical Mission Societies drawn from many countries in Europe. He himself spent several years travelling to and living in St. Petersburg, especially during the Winters, while preaching and teaching Bible study classes. This work was directed particularly at the Russian people.

During World War I he began to teach and distribute Bibles among the Russian prisoners of war in Germany and soon also among the Russian refugees who were primarily women and children. It was at this point that this evangelizing effort was expanded to include a relief effort for many of these needy people. This organization, called Light in the East (*Licht im Osten*), was headed by Kroeker who was well known and knew much about the people of Russia. Thus when famine in the 1930's raised its ugly head again, he, because of his knowledge and his contacts in Russia, was in an extraordinary position to help both with relief and by sending religious literature to many parts of the Soviet Union. He died 12 December 1948 in Germany. Since PJ 1883 had been a prisoner of war in Germany while serving in the Russian medical corps during WWI, it is likely that they met

[29] The details of this man's life are taken from a memoir written by his daughter, Maria Kroeker, *Ein reiches Leben* (Weinsburg, 1949). I thank Dr. John B. Toews for lending me this book.

at that time. Whether JP 1910 ever made use of this organization I do not know. From later correspondence it seems that he sent mainly money to the family by registered mail. The family does seem to have managed to get some food for there is a reference to painting easter eggs, as well as the mention of the baking of bread. Clearly they still had some of the chickens which they had so carefully raised in the previous year. However, in March of this year at least in some villages a tax of two chickens out of every ten had been imposed on the villagers. So even this source of food was being squeezed from them. The starch that ASR 1889 used to bake bread was probably an allotment earned by working in the Artel.

Brief

1931 [warscheinlich um Ostern]

Ich bang mich so sehr

Lieber Jasch,[ii] bin schön gesund. Brot schmeckt schön. Milch trink ich sehr gerne. Hatte auch Eier gefärbt. Ich hatte 5 Eier rot, blau, grün und braun. Ich hab auch ein Häschen. Ich bang mich so sehr, komm hier her. Waren gestern zu Tante Anna spazieren gegangen. Ich kann schon sehr laufen und spielen, spiel auch blinde Kuh. Kann auch schon abzählen. Eins [*uzw.*] Kann auch schon singen, auch schlimm sein. Deine Schwester Änni

[ii] Diese nächste zwei Briefe hat wahrscheinlich die Mama, ASR 1889, fur die Kleinen aufgeschrieben.

Letter

1931 [around Easter][30]

I Miss you so Much

Dear Jasch [Jacob]: Am very healthy. Bread tastes good. I very much enjoy drinking milk. I had also colored eggs. I had 5 eggs, red, blue, green and brown. I also have a rabbit. I miss you so much, come here to us. Went to visit Aunt Anna yesterday.[31] I am already able to run and play well, can also play Blind Man's Bluff. Can also count already, one, two. etc. Can also sing nicely, also be cantankerous. Your sister Anni

[30] The next two short letters were written for the youngest girls by ASR 1889 since they were not yet in school and could not write. Clearly some of the editorializing is not of their own thinking.

[31] This was Anna J. Rahn 1884, one of PJ 1883's sisters, who never married and lived into her nineties. She lived in Sparrau with her half brother, Jacob Jacob Rahn 1875, and her sister, Helena (Rahn) Dueck 1891 and their families in the house originally owned by their father, Jacob Johann Rahn 1840, my great grandfather. Anna seemed to visit in Kontiniusfeld quite often according to my aunts.

Brief

1931 [warscheinlich um Ostern]
Ich will nicht im Kindergarten

Auch ich komm zu Dir. Weißt Du ich und Justa hatten 6 Eier. Färbten die mit Liese Derksen zusammen. Es hat Ostern gut gegangen. Ich liebe immer das Toben. Ich will nicht im Kindergarten hinein. Ich will bei Mama bleiben. Ich wünsche mir schon schönes Wetter zum draußen spielen. Drin soll ich nicht toben. Drin schelten sie mich immer. Wir hatten auch Gerste gesät zu Ostern. Jetzt hat das Kaninchen die Gerste abgerupft. Deine Schwester Mika.

Letter

1931 [around Easter]
I don't Want to go to Kindergarten

I also am coming to you. Did you know that I and Justa had 6 eggs. Painted them together with Liese Derksen. Easter went really well. I enjoy the romping around. I don't want to go to Kindergarten. I want to stay here with Mama. I am already longing for nice weather so that I can play outside. Inside I am not allowed to romp around. Inside they are always scolding. We seeded barley [be]fore Easter. Now the rabbit has snipped off the shoots. Your sister Mika.

COMMENTARY

In February of 1931 a large push was made to set up childcare and kindergartens in the Mennonite villages so that *the women can participate more directly in the work of the collective*. Moreover children were to remain in school until the age of fifteen.[32] The preceding letters, notes and postscripts reflect the changes in daily routine being forced on the villagers. Aganetha Schartner Rahn repeats how busy everyone is, that even the girls have to work in the fields. The collective kitchen is one more device by which the villagers are incited to work in the Artels. In March there had again been searches for hidden flour. Wheat was extremely costly.[33] In the midst of this it seems that members of the family all worked hard in the Artel to receive whatever food would be handed out. Moreover, they seem not to have defied the authorities, for frequently in some letters we read that there had been a search in the village, but *at our house nothing* [contraband] *was found*.

Yet, the letters express the extreme frustration brought on by all these changes. Mika does not want to be separated from her mother by the new school regulations. Both PJ 1883 and ASR 1889 are frustrated by the attempts at mechanizing the agricultural process. The lack of workers in seeding the crops is the direct result of the dekulakization process in which many families had been removed from the village and thereby the manpower in the village had been reduced. The collectivizing of all the village horses and their use for the meat supply, and the insistence of the authorities on using only tractors, that are to make up for the lack of horses, for the plowing according to a preordained schedule is not succeeding because of the condition of the tractors. Finally, the constant physical hunger resulting from insufficient provisions for all the hard work give a sense of hopelessness and depression. They are doing their best to contribute, but they are still losing ground.

[32] DB 11/03/1931, p. 5; MR 20/05/1931, p. 7.

[33] DB 22/04/1931, p. 4.

Brief

1931[iii]

Schick mir eine ...

Lieber Bruder, ich will Dir mal einen Brief schreiben. Bin schön gesund und wünsche Dir dasselbe. Ich gehe alle Tage zur Schule. Sitze bei Liese Dirk. und Käte Schartner. Fangen auch schon an Handarbeit zu machen. Bald ist Weihnachten, bitte schicke mir eine Harfe?? Deine Schwester Justa.

Bitte studieren!

Jasch komm hus. Bling mie en bild ??? Lene/ schw. Anni. Eck bang mie so um die. Komm hol uns uck he. Scheck mi uk nu ein Bild. diene Schwasta Mika. Lena sagt du sollst deine photographie schicken. Anni will auch eine bitte haben. L.[ena]

iii [warscheinlich auch um Ostern.] Mama [ASR 1889] hat die Briefe für Anni und Mika mit ihren Brief aufgeschrieben. Seid Justa schon in der Schule war hat sie natürlich selbst ihren eignen Brief vorher geschrieben. Dann hat Schwester Lena [Rahn Pauls 1914] auf der Rückseite im Platdeutschen für die zwei Kleinsten was da folgt aufgeschrieben. Ich glaub diese kleine Briefe waren vor Ostern geschrieben, und dann als die Kleinen die Bilder von JP 1910 erhielten bedankten sie sich in die Briefe die Mama aufgeschrieben hat.

Letter

1931[34]

Send me a...

Dear Brother, Well I want to write you a letter. Am nice and healthy and wish you the same. I go to school every day. Sit beside Liese Dirk.[sen] and Kate Schartner. Have also already begun to do handicrafts. Pretty soon it will be Christmas, please send me a harp. Your sister Justa

Please Study!

Jake, come home. Bring me a picture[?]. Lena/Schw. Anni. I miss you so much. Come and take us away [to where you are]. Now also send me a picture. Your sister Mika. Lena says you are supposed to send your photography. Anni also wants to have one please. L.[ena]

34 This letter may well have been written on April 11 before those of her sisters written for them by mother [ASR 1889] at Easter time. The reference to the harp may be a mistranslation. The original is very diffcult to read. On the back of Justa's letter Lena (Rahn) Pauls 1914 wrote for the other two girls, Mika and Anni, in low German except for the first and last two lines. Since these letters request photographs, and the other letters written by Mama on the 16th thank them for the photos, it is likely that they received the photographs on Good Friday as a kind of Easter present after having written these letters beforehand, without knowing the pictures were already on the way.

Brief

1931 04 23

P.[apas] Bru. Hein ist gestorben

Da wir heute einen Brief von Dir erhalten haben, so will ich auch gleich versuchen, den zu beantworten. Der Brief ist vom 31. März, hat nicht lange gegangen. Wünschen Dir die beste Gesundheit und das beste Wohlergehen an Leib und Seele. Gesund sind wir, dem Herrn sei Dank, bis jetzt noch alle. Nur Mi.[ka] ging heute früh schlafen, brennt heiß. Ich weiß nicht, ist das vom Barfuss laufen? Sie sind noch nur 2 Mal barfuss gegangen. Das wird noch gar nicht so schön. Es hat Gestern und Heute schon mit Erde gestürmt. Aber jetzt ist es still geworden. Es war großer Sturm. Es ist Abend. Die Kleinen drin schlafen ohne Sorgen. Le.[na] ging noch nach Sp. zu A.[nna] R.[ahn] nach Blumensamen und gelbe R.[üben] Samen. Wenn es geht, wollen wir Morgen Kartoffeln setzen. Heute haben wir schon mehrere gesetzt. Ha.[ns] ist Abend nach Hause gekommen. Hat sich auch schon die Neugierde gestillt. P.[apa] ist jetzt schon die 5. Nacht auf der Steppe. Muss bei den Traktoren nach den Pflügon schauen, ob das fährt. H.[einrich] war vorige Nacht zu Hause, musste aber Uhr 4 wieder gehen. 5 Traktoren gehen, leider bleiben sie oft

Letter

1931 04 23

P.[apa's] brother Hein has died

Since we received a letter from you today, I will also try immediately to reply. The letter was from the 31st of March. Did not take long [to arrive]. Wish you the best of health and well being in body and soul. Up to the present, thank God, we are all healthy. Only Mi.[ka] went to bed early tonight, had a fever. I am not sure, is it the result of running around barefoot? They [the little girls, Anni and Mika] have only been [outside] barefoot twice. The weather isn't really that nice [yet]. Yesterday and today we had dust storms, but now it has grown quiet. It was a huge storm. It is evening. The little ones are asleep within without any worries. However, Le[na] has still gone to Sp[arrau] to A. R.[35] for flower seeds and some yellow beet seed. If it is possible, we want to plant potatoes tomorrow. We already planted several today. Ha.[ns] has come home this evening. He seems to have satisfied his curiosity. P.[apa] is now spending the 5th night on the field. Is supposed keep watch over the plow (*Pfuegon*) of [hitched to] the tractor, whether it is running [working]. H.[einrich] was home the night before last, but had to leave again at 4:00 [AM]. 5 tractors are running. Sadly, they often come to a standstill. The [system]

[35] Anna Jacob Rahn, 1884, sister to PJ 1883.

stehen. Es ist einmal so. Es sind 2 Partien, eine P. am Tage und eine in der Nacht. Hier werden sie Morgen vielleicht beendigt mit den Traktoren pflügen, dann gehen die nach Sp. H.[einrich] wird vielleicht einen Anzug bekommen als Traktorist. Die sehen auch gräulich aus, wenn sie von der Steppe kommen. H.[einrich] hat diese Woche noch immer aus der Küche gegessen. Nur gestern Abendbrot aß er zu Hause. Heißt da M.[ittagessen] und Abendb.[rot]. Ach, da gibt's manches, der Eine isst, der 2. bekommt nichts und dann wird gescholten, es geht so. Du hast keine Vorstellung von dem Leben hier, so schwer wie die Lage auch ist und traurig und wenn man dann auf das Leben der Unsrigen schaut, ach wie traurig, wie wird uns erlaubt, was für kein Unrecht eingestechen wird. Meinst Du wir würden nicht besser leben können, wenn nicht Sünde wäre, aber möchte der liebe Gott uns vor allem Unrecht bewahren. Er hat es bis jetzt noch bewiesen, dass er uns noch nicht vergessen hat. Haben heute das Öl erhalten, bekommen 2 Pfund weniger als ein Pud. Endlich auch erhalten. Freuen uns sehr dazu. Ä.[nni] sagte, *jetzt haben wir Öl, jetzt haben wir alles, das macht alles der liebe Heiland*. Ich musste jetzt Abends Allen den Brief von Dir vorlesen, dann hören sie zu.

is as follows. There are two work parties, one for the daytime and one for the night. Here they will perhaps finish plowing with the tractors tomorrow morning, then they will move on to Sp.[arrau]. H. [einrich] as a tractorist[36] perhaps will get a uniform.

They [the tractor drivers] all have a gray appearance when they return from the fields. H.[einrich] this week has always eaten at the [collective] kitchen. He only ate yesterday's supper here at home. That means [provides] midday and supper meals. Alas there is many an issue; one person eats, the second one gets nothing, and then there is great complaining. That is how it is! You cannot imagine what life is like here, how difficult our situation is, and how sad. When one considers the lives of our people it is really sad [to see] what is allowed to us, [while another] is incarcerated for the same not unjust deed. Don't you think we would live better if there was no sin, but may our loving God protect us from injustice. He has until now demonstrated that He has not yet forgotten us. Received oil [cooking] today, got 2 pounds less than a Pud.[37] Finally also received it. Are very happy for it. A.[nni] said, "Now we have oil, now we have everything. Our loving saviour does all that!" I am required to read out loud the letter from you. Then they all listen. But the first

[36] [technician, driver] He had applied in the autumn for work in the MTS. See letter 1930/12/28 and commentary to letter 1931/03/02.

[37] Approx. 36 lbs., so amount of oil was about 34 lbs.

Aber das erste fragen sie, ob auch mal an sie geschrieben ist. Die beiden kleinen gehen oft Spazieren zu den Nachbarn. Wir backten heute Pfannkuchen, den haben die Kinder gegessen. Morgen will ich mal Bohnensuppe mit Kartoffeln kochen. So kamen sie in der Küche. P.[apa] hatte die geschmeckt und die schmeckt schön. Du fragst nach Hü.[hner] und E.[ier] u. Gartenflük. Auf dem Gartenflük haben wir R.[üben] gesät. Aber hier unten wo das Stroh war, wollen wir Kartoffeln setzen. Dann bekommen wir noch auf der Steppe zu Kartoffeln. Hühner haben wir aber da müssen wir für jedes Huhn 30 Eier liefern. Enten haben wir bis jetzt noch vier und die haben schon 19 Eier gelegt. Eine Glucke brütet schon. Grüße alle die im Hause sind. Die Fürbitte tut not.[iv] Wenn jetzt eine brütet, dann will ich auf Enteneier nur setzen. Lena kam Abend nach Hause und erzählt, dass P.[apas] Bru. Hein gestorben ist, die zogen vor paar Wochen dahin, wo der Br.[uder] Da.[vid] R.[ahn] wohnt. Sie

[iv] Die nächsten Zeilen sind als Nachschrift an geschrieben auf Seite 3.

thing they always ask is whether finally they have been written to. The two smallest ones often go visiting at the neighbours. We baked pancakes today. Those the children devoured. Tomorrow I want to serve bean soup with potatoes for a change. As a result the children came into the kitchen. P.[apa] had tasted it and it tasted good. You ask about chickens and E.[ggs] and about the garden plot. On the garden plot we have seeded R.[38] But at the lower end near us, where the straw had lain, we want to plant potatoes. In addition we will receive an allotment on the field (Steppe) for potatoes. Chickens we have, but for every chicken we have to deliver 30 eggs.[39] To this point we have four ducks and they have laid 19 eggs. One hen is already brooding. Greet all who reside in [your] house. Intercession is needed![40] If now another [hen] broods, then I want to set her only on duck eggs. Lena has just arrived home and recounts that P.[apa's] brother Hein has died. Only a few weeks ago they moved to the area where brother David R.[ahn] lives.[41] They had

[38] Probably *Rueben*, beets.

[39] [to the communist authorities, as a chicken tax.] The nature of this chicken tax seems to vary from village to village and from time to time. E.g., see DB 25/03/1931, p. 4.

[40] The following is the third page of the letter, added on later the same evening. The back [p. 4] is separately dated the 24/04.

[41] Heinrich Jacob Rahn, 1872 was the oldest brother, born of the first wife, Tina Vogt. David Jacob Rahn 1878, the first born of the second wife, Helena Dueck 1849, lived in the Caucasus.

ließen alles stehen und fuhren still ab. 2 Jungen waren da. Jetzt kam Peter und wollte die Sachen abschicken und in der Zeit ist er gestorben. Es wird vielleicht 3 Wochen sein, dass sie da sind und jetzt ist er tot. P.[apas] Schwager, Tante Greta ihr Mann, sitzt auch, haben auch nichts mehr. Sie wohnt bei ihrer Tochter. Die anderen leben ja noch alle. Haben auch viel zu sorgen um das tägliche Brot. Wir wollen was Öl verkaufen, um Brot zu kaufen. Das Öl ist sehr teuer, ist nicht zu bekommen. Ohm H. kommt alle Tage bei uns, mitunter bis zwei Mal am Tage. Er sagt, er ist schon gesund. Er schalt Heute schon, er wollte auch Öl haben und bekam nicht, sie hatten ihm ja auch Samen weg genommen. Aber die bekamen auch noch alle. Als er krank war, musste ich ihm Mus kochen von Buttermilch. Die schmeckt ihm, bestellte sich Heute auch wieder die Buttermilch, denn seine Lage ist auch schwer. Nun gute Nacht, vielleicht schreibt P.[apa] auch noch ein bisschen. Lena bestellt zu grüßen. Wir haben

simply left everything as it was and drove off quietly. 2 Boys were staying there. Now Peter[42] returned and wanted to send their possessions away [to them], but in that time he [HJ1872] died. It is perhaps 3 weeks that they were there and now he is dead. Papa's brother-in-law,[43] Aunt Greta's husband, is also sitting [in jail], also have nothing anymore. She is living with her daughter. The others [relatives] are still all alive. Have many concerns about their daily bread. We want to sell some oil in order to buy some bread.[44] Oil is very expensive, is impossible to obtain. Ohm Hein [Janz] comes to our house every day, sometimes twice per day. He says he is nice and healthy. He was already complaining today; he wanted to have some oil[45] but did not get any. After all, they had also taken seed away from him, but they[46] still all got some. When he was sick, I had to cook a moos with buttermilk. He loves that taste, ordered some more buttermilk for himself, for his situation is very serious. Well, good night. Perhaps P.[apa] will also write a little bit. Lena sends her greetings. We have often

[42] Probably one of HJ 1872's sons.

[43] Margereta Jacob Rahn 1880 married a Kornelius Friesen. This is confirmed by JP 1910's paraphrase of this part of the letter in his memoirs. They lived in Sparrau and had two daughters, Anna and Maria Friesen.

[44] [Probably means flour, *Brot*].

[45] [Presumably from the communist authorities].

[46] [The local communist leadership].

oft gesagt, sie soll schreiben, aber Du solltest erst schreiben, sagte sie. [ASR 1889]

24. A. 1931[v] Haben eben Mittag gegessen, Linsensuppe mit Kart-offeln. Ohm H.[einrich] aß auch, er sagte hat gut geschmeckt. Ist heute schon 3 Mal hier gewesen. Hat sich auch Öl geholt. Es ist heute wieder Sturm. Die Traktoren gingen heute schon zum nächsten Dorf. Dann darf P.[apa] nicht mehr des Nachts. Du fragst nach etliche hier, der in Großpapas Haus hineinzog. Der ist in allem der Vorderste, so dass es bei manchen zu kämpfen gibt, aber Alles muss schweigen. P. W. leben nach alter Art. J. U. denen geht es nur schlecht. Öl bekommen sie keines, Brot sehr knapp. Muss auch schon kaufen. Im Ar.[tel] dürfen die Kinder nicht arbeiten. Weil er P.[rediger] ist. Die anderen sind alle in Arb.[eit] sogar Ohm H. Nun werde schließen. Vormittag war

[v] Den nächsten Tag noch zugeschrieben weil die vierte Seite noch leer war.

encouraged her to write, but she says you are to write [to her] first. [ASR 1889]

1931 04 24[47] Have just eaten the noon meal, Lentil soup with potatoes. Ohm H[ein] ate as well, he said it tasted good. Has been here 3 times already today. He came and got some oil. Today again there is a storm.[48] The tractors all went to the next village today. Then Papa is not required to work at night. You ask about some here, about the one who has moved into Grandpa's[49] house. In everything he is the very first. So many a one has some serious struggles, but everything [everyone] must remain quiet. P. W.[50] live in their previous manner, J. U.[51] are only faring badly. Oil they receive not at all, bread only very sparingly. In the Ar.[tel] the children are not allowed to work, because he [the father] is a P.[reacher]. Everybody else is at [allowed to] work, even Ohm Hein. Will end

[47] Written on the back of 1931 04 23, p. 3, which was still empty at this point.

[48] These were serious wind storms that created huge dust storms by blowing the topsoil around and made it very difficult to seed the fields.

[49] [Probably B. B. Janz 1852's house]. JP 1910, in his paraphrase of this sentence, simply calls him *the chief of the village*, but does not name him. According to my aunts, one of the main Communist activists was a Peter Gulig, the son of the former village cowherds. On a later map of Kontiniusfeld drawn by Maria Kaspar, a Johann Dueck lived there during some part of the 1930s.

[50] Perhaps Peter Willms, who lived in Kontiniusfeld.

[51] A preacher John Unrau, according to JP 1910's memoirs.

Tante David Dirk. hier, holte sich Milch denn ihre Annchen liegt auch krank. Ohm H. liegt auch krank. Hat. Schnupfen. P.[apa] schläft, ist müde von der Steppe. Hein fuhr zum nächsten Dorf. Ha[ns] ist auf der Steppe. Le[na] schafft im Garten. Die Kleinen 3 gingen zu Töw. Grüße aus die Ferne. Glaube alles beantwortet zu haben. [ASR 1889]

now. Before noon Mrs. Dav.[id] Dirk[sen] was here, got some milk because her little Annchen is lying sick. Ohm Hein is also sick. Has the sniffles. P.[apa] is sleeping, is worn out by [work in] the fields (*Steppe*). Hein [HP 1911] drove to the next village. Ha[ns] is out in the field (*Steppe*). Le[na] is working in the garden. The 3 little ones went to Toews. Greetings from far away. Believe I've answered everything. [ASR 1889]

COMMENTARY

As a result of the collectivization process different types of agricultural organization seem to have emerged over time. The most dictatorial or hierarchical in nature was the State farm or so-called *sovkhozy* operated by the state with hired labour. While travelling through the Ukraine in 2007 we came upon one of these that still seemed to be in operation and served more as a place for mentally challenged farm labourers than anyone else. The second form of collectivized peasantry was the *kolkhoz*. It was intended to be the basic means of collectivizing and dekulakizing the agricultural industry. In the strictest *kolkhoz* there was common ownership of everything, land, houses, cattle, food and all consumer goods. However at the end of 1931 almost forty percent of the peasants in the Soviet Union had not yet been fully collectivized inside the *kolkhoz*.[52] The *artel* seemed a slight compromise between the *kolkhoz* and traditional village life. In it the villagers, who had not yet been disenfranchised, continued to live in their own houses, at first had their own cow and horse and a small garden plot with which they could do as they wished. The rest of the land as well as the properties of the banished kulaks were merged and worked collectively as the property of the kolkhoz.

The artel seems to have been the primary mode of organization in Kontiniusfeld, although officially it had been designated *Kolkhozy Stalina*.[53] Of course, there was continuous pressure, especially in the Spring of 1931, to push the village into a more severe form of the kolkhoz through the introduction of collective kitchens, daycares and kindergartens [where the children were fed at least the noon day meal], the confiscation of the individual cows and horses, and the provision of some food rations from the collectivized store. The collectivized property, of course, required some one to work it in order for it to provide the required quota of grain demanded by the state from each kolkhoz. The villager provided the physical labour and the regional MTS [tractor brigade] was expected to furnish the machinery

[52] Sheila Fitzpatrick, *Stalins's Peasants* (1994) p. 7, 8, 153.

[53] I am grateful to Colin Neufeldt for this information drawn from the State archive in Zaporizhzhia, Ukraine. Sparrau was called *Politotdelez*.

to seed the fields and reap the harvest. In return the local kolkhoz had to pay for this help in kind [mainly in grain]. Thus the villagers, as members of the artel, were the work detail for this collectivized effort to produce enough grain to fulfil the state set quota and still have enough left to feed themselves. However, the extreme demands of the state made it virtually impossible for the members of the artel to provide adequately for themselves and their families.

In some measure the return on their labour was also affected by the demands of the local bureaucracy. In his book[54] describing the collectivization effort in his Ukrainian village in 1930, Miron Dolot [*pseudn.*] finds that in a village of 4000 people, the administration of the collectivized agrarian system required more than 496 people to ensure its functioning. Jakob Neufeldt[55] estimates that in Gnadenfeld the cost of using the MTS to plant and harvest the crops, cost the collective twenty-five percent of the local harvest.

Insight into the experience of the Mennonite villages [particularly in the Molotschna] may be gained from a description of the Artel's structure dated 7 December 1930.[56] In this unnamed village, the actual head was a communist assisted by two labour leaders and three people elected from the village to serve as a local village council. In the office three book keepers and several brigadiers supervised the daily activities of the artel.[57] The village workers were divided into male and female brigades headed by one of those from the office. Each brigadier kept track of the hours worked by the people of his brigade.[58] The work was divided into five categories and people received credits based on the type [or category] of work and the

[54] *Execution by Hunger* (1985), p. 9.

[55] *Tiefenwege,* p. 24-25.

[56] MR 1931/02/25, p. 6, 7. Although the village is unnamed, the writer refers to persons from Kontiniusfeld and Gnadenfeld by name, and so is not unfamiliar with the region. For another description of how the system has deteriorated, see DB 1932/03/02, p. 3.

[57] No wonder Ohm Hein complained that while *they* all [all 14 or 15 of them] got cooking oil he got none even though he had surrendered sunflower seed to the state.

[58] In Kontiniusfeld there were two permanent brigades, though there may have been others for ad hoc tasks. JusP 1922, *Mein Wegbegleiter Jesus Christus,* p. 5, 15.

number of hours worked. The normal working day was ten hours and anyone who worked longer received time and a half for the extra hours. Most of this work was connected with the seeding, cultivation and harvest of the crops. The entire harvest including the potatoes and the sunflower seeds was confiscated by the state and then individual participants were allowed to buy back some small portions for seeding in the Spring at excessively high fixed prices with money earned by working in the artel. No wonder the writer of the description of the Artel echoes almost *verbatim* the words of PJ 1883, *what good is all that working, if in return for all that work one has to starve for the whole year, for they take everything away.*[59]

Throughout the remainder of 1931, and all of 1932 very little [or, perhaps, no] correspondence seems to have been received directly by JP 1910. The reasons for this are probably several. First, a new wave of terror, arrests, exiles and incarcerations began in May 1931 throughout the Molotschna. A large percentage of the Mennonite population of Halbstatt [about 1500 people] was evacuated,[60] some probably to the newly founded resettlement villages of Octoberfeld and Neuhof, others to Siberia or other Northern areas of Russia to be industrialized. Though some of these exiles were allowed to return to their home villages later, at that moment this event must have had a terrifying effect on the surrounding villages. In addition, a number of men from other Mennonite villages were arrested on the night of May 17, 18 and jailed in Halbstatt, and were still awaiting sentence in June.[61] In June of this year some 25,000 Ukrainians, Germans and Bulgars living to the South of the Molotschna colony were rounded up and shipped to the North from the Klippenfeld/ Stuljnevo railway station. The transportation of such a mass through the Eastern end of the Molotschna did not occur with out attracting attention and instilling fear into the lives of those in the Mennonite villages, some of whom were included in this exile.[62] The need

[59] Letter in 1931/04/10.

[60] DB 10/06/1931, p. 4; DB 6/05/1931, p. 4.

[61] DB 19/08/1931, p. 5.

[62] DB 29/07/1931, p. 4; 23/09/1931; MR 21/10/1931. Whether 25,000 was the actual number is difficult to confirm. What is important here is that this number originates with a Mennonite writing from the Eastern end of the Molotschna at this time. So it is clear that this exodus made a major impact on the Mennonites observing it.

to harvest the crops brought a halt to this reshuffling of the population, but in November some seventy-five Mennonites from the Molotschna region [frequently associated with the grain trade] were again arrested and jailed. Some of the resettlement villages were emptied of their inhabitants who were sent into the Urals.[63]

Second, amid all these arrests and incursions into the private lives of the villagers, the second half of 1931 found the people of the Ukraine under severe pressure to hand over all foreign currency and gold that they might have in their possession. Beginning with the arrests in May, and increasingly in the latter part of the year, homes were searched for gold or dollars and people arrested, if the authorities were aware of any contact or correspondence with foreigners, in an attempt to boost the foreign currency reserves of the state. In some cases, people were jailed and then promised freedom if they handed over dollars.[64] In the Chortitza-Rosenthal settlement in the early autumn people had been promised shoes and galoshes, if they made deposits of money. On the contrary none ever materialized but the money was never returned.[65] This wide spread attempt to expand by any and every means the foreign currency reserves of the Soviet regime extended on well beyond 1932 as we shall see later.

Third, the harvesting of the crops in the Molotschna in 1931 was a complicated and time-consuming business. Estimates about the nature and quality of the harvest vary according to the source and the area. However, the general consensus about the harvest in the Molotschna in 1931 is that the grain crops were very good;[66] in Ohrloff the crop was described as the best since the Soviets took over. After the threshing, all of the grain was immediately hauled off to the train station.[67] There it sometimes lay on the ground rotting, because the transports could not keep up with the volume

[63] DB 6/01/1932, p. 4; MR 23/12/1931, p. 7.
[64] DB 6/01/1932, p. 4; MR 24/02/1932, p. 7.
[65] DB 16/12/1931, p. 6.
[66] DB 23/09/1931, p. 4; 11/11/1931, p. 4; 2/12/1931, p. 2; re Ohrloff, MR 24/02/1932, p. 7.
[67] DB 9/09/1931, p. 4; 12/08/1931, p. 4.

of the harvest.[68] Moreover, the threshing had still not been completed by the end of October[69] and beets, sunflowers and corn (*kukurus*) had not been cleanly collected from the fields by the end of the year.[70] Colin Neufeldt estimates that thirty percent of the harvest was lost and that in spite of all the pressure that was put on the agricultural population, the Ukraine failed to deliver its required grain quota in 1931.[71] This occurred in spite of the fact that in the artels the day of rest was cancelled during the harvest season, although the five-day work week with a day of rest in between had been in effect since the spring. Even children were being put to work.[72]

Fourth, for almost the entire year people, especially the children, had been living on the edge of starvation, whether they worked in the Artels or not.[73] Even during the harvest season efforts were made to ensure that every kernel of grain was confiscated by the state. For instance, in some villages the private use of white flour (grain flour) had been forbidden.[74] Even in the collective kitchens there was a rather arbitrary distribution of food.[75] Finally, in November some rye flour was distributed, but only to workers in the Artel or on the Steppes.[76] In a letter dated 24/12/1931[77] the writer outlines what supplies of food they [a couple] have left to see them through the winter, 200 grams of bread per day via ration cards: approximately 1350 kilograms of potatoes, 72.5 kilograms of beans, 1 litre of sunflower oil, 10 to 15 cabbages, some beets and carrots and several bags of porridge, with little hope of getting anything else. Under these conditions, not only

[68] MR 23/12/1931, p. 7; DB 2/12/1931, p. 2.
[69] MR 23/12/1931, p. 7, DB 2/03/1932, p. 3.
[70] MR 24/02/1932, p. 7.
[71] Neufeldt 1999, p. 238.
[72] MR 23/12/1931, p. 7, DB 1/12/1931, p. 2. By the Fall of 1933 the authorities in the Crimea and probably elsewhere seem to have returned to the seven-day work week. DB 14/03/1934, p. 4.
[73] See letters 1931/03/05; 04/10; DB 29/07/1931, p. 4.
[74] MR 23/12/1931, p. 7.
[75] See letter 1931/04/10.
[76] DB 2/12/1931, p. 2.
[77] DB 10/02/1932, p. 5.

were the villagers required to supply the needed labour, but, in order to ensure their own survival they needed to tend their own vegetable gardens and domestic fowl.[78] At the same time they had to be vigilant least thieves should steal what little they still had.[79]

Finally, the anti-religious campaign begun earlier in the year intensified. Whereas earlier mainly the religious leadership and the teachers had been targeted, now the rest of the population also began to experience pressure. For instance, in October a proclamation was circulated, at least in some villages, that anyone who failed to join the *circle of the godless* would lose their ration cards.[80] At Christmas time in Halbstatt, and in some other villages as well,[81] workers who attended celebrations were docked part of their wages. Children who attended church rather than the required anti-Christmas rallies were fined five to eight Rubles and people, who had set up Christmas trees, thirty Rubels.

At some point in this year letters from the Rahn family in Kontiniusfeld stopped arriving in Canada and no regular correspondence was re-established until December 1933. The Aunts[82] only explanation was that they were not allowed to write. Yet it is clear that correspondence from Mennonites in the Ukraine published in the *Mennonitische Rundschau* and the *Bote* did arrive in Canada during these years. Given the horrendous, incompetent and arbitrary behaviour and the hostility to religious faith of the Communist authorities, additional reasons for not writing might be that, first, the family did not wish to draw attention to themselves by engaging in writing to foreign addresses. Second, they were utterly exhausted from working in the artel and using every free moment to tend and keep watch over their own

[78] See letter 1931/04/23.

[79] Clearly the increase in lawlessness was engendered by the continual starvation. E.g., in the forests near Halbstatt a band of thieves had taken up residence and threatened many of the villages or travellers. DB 18/03/1932, p. 5.

[80] DB 2/12/1931, p. 2.

[81] MR 24/02/1932, p. 7; DB 9/03/1932, p. 3.

[82] Lena Rahn Pauls 1914 and Anni Rahn Wiens 1927.

private gardens and domestic fowl.[83] Although JP 1910 was sending them money[84] [and perhaps food such as rice] clearly things did not get better for them throughout the latter half of the year. Finally, several letters of 1931 [01/22, 03/31, 04/23] indicate that writing materials [and perhaps money for postage] had run out and no more were available.

[83] See letter 1931/04/23.

[84] See letter in later chapter 1934/03/29.

1932
in Kontiniusfeld

Family name abbreviations used in discussion of the letters

ASR 1889	Aganetha Schartner Rahn 1889 – 1968
PJ1883	Peter Jacob Rahn 1883 – 1942
JP1910	Jacob Peter Rahn 1910 – 1991
HP 1911	Henry Peter Rahn 1911 – 1996
JohP 1913	Johann (Hans) Peter Rahn 1913 – 1944
HelP 1914	Helena Peter (Rahn) Pauls b. 1914
JusP 1922	Justa Peter (Rahn) Prei Geddart 1922 -2004
Anni P 1927	Anna Peter (Rahn) Wiens b. 1927

Kontiniusfeld 1932

Although no letters survive from 1932, for the sake of completeness and context it is necessary to summarize briefly the main events and the nature of our relatives' existence during that year and the following one. After the terror of 1931, the next year did not begin well and only got worse. As already pointed out, aside from the arrests, ejections, exiles and deportations of individuals and families from Kontiniusfeld/Sparrau in 1931, partial starvation was a constant throughout much of that year. The authorities continued to seize grain and flour throughout the year, and when the grain harvest came they made sure that virtually none of it fell into the hands of the local villagers. What maintained the PJ Rahn family in 1931 was their industriousness to provide their fair share of labour to the state and then to ensure their own survival through whatever private means remained open to them. The mother, Aganetha Schartner Rahn [1889], is seen from the letters to have been a hardworking, enterprising woman who cared deeply about the children of her blended family. She husbanded the hatching and growth of flocks of chickens and ducks, and planted and weeded their private vegetable gardens to the full extent that was allowed. The older children, Heinrich, Hans, Lena and even Justa, participated actively in the domestic chores, and so contributed significantly to the survival of the family. In addition, Heinrich, by working in the MTS, often received meals from the state or small amounts of raw foodstuffs that he shared with the family. In spite of some trepidation on the part of the parents, Hans began to maintain and trade for rabbits, which must have proved another source of food in the fast diminishing circumstances. Peter Jacob Rahn [1883], in addition to working in the fields of the Kolkhoz, was literate and therefore did some work in the local artel office from time to time.[1] He also tended their land allotment on the steppes, which was planted with corn, sunflowers and watermelons, while

[1] Mentioned by JP 1910 in his personal memoirs and paraphrases of some of the letters of 1930/31.

potatoes and beets were planted close to the house in their personal garden. Sometimes this private work was done with other relatives of the family. PJ1883 also seems to have been something of a support and confidante to other members of the village who might wish to talk about their religious faith in the midst of this terrible experience. Thus the family's own summer activity probably helped them through the winter of 1931/32.

Although there was certainly starvation in the villages of the Eastern Molotschna, in the nearby village of Gnadenfeld[2] only a few people died in the winter of 1931/32, a fact that suggests that similar conditions probably reigned in Kontiniusfeld/Sparrau, for it was only about four kilometres away. The survival of most of the population resulted partly from the failure to harvest the fields cleanly in the autumn of 1931, which meant that cornstalks and sunflowers as well as some ears of grain were left on the fields of the kolkhoz. During the winter months people were constantly searching these frozen remains for kernels, seeds and grains.[3] This practice proliferated throughout the year, even in very rich settlements,[4] until in the autumn of 1932 the government decreed the gleaning of the leftovers from the fields of the kolkhoz *theft of state property punishable by death*. The fields were then guarded by watchmen.[5] Thus from March 1932 onward the severe famine increased throughout the Ukraine, even in the artels.[6]

Not only human beings but also farm animals suffered severe depredation. Horses were being killed for food, livestock was weak and diseased since even their food was being stolen, and many newborn calves and foals did not survive.[7] By mid-year many cattle, pigs and sheep were dying because of lack of food. Many of these carcasses were being collected, cooked up and fed to the pigs in the collective barns or even eaten by the starving villagers

[2] MR 7/12/1932, p. 7. Jacob A. Neufeld, *Tiefenwege*, p. 14, who was a resident of Gnadenfeld, says that three people died there of starvation.

[3] MR 3/08/1932, p. 6, 13/07/1932, p. 6; DB 21/12/1932, p. 3.

[4] MR 6/07/1932, p. 7, 28/09/1932, p. 6-7.

[5] DB 2/11/1932, p. 4, 21/12/1932, p. 3.

[6] MR 16/03/1932, p. 7, 20/04/1932, p. 7, 25/05/1932, p. 6, 1/06/1932, p. 7, 21/09 1932, p. 7; DB 18/05/1932, p. 4, 10/08/1932, p. 4, 7/12/1932, p. 4.

[7] DB 2/03/1932, p. 3.

themselves.[8] In most villages the cows seem to have been collectivized by March but even some of these were stolen.[9] In June it was announced that those villagers who worked in the artels would again be entitled to keep a cow. But this seems to have been more show than reality, for this partially reduced the demands for nourishment on the local authority since people themselves could supposedly now nourish their families. But there was also no feed available. As a result many of the cows did not produce much milk[10] or died or were sent to the cities for food. At least in one village there were only ten cows available for 575 people.[11] During the winter months the chickens stopped laying eggs because of lack of proper feed, and most of those that were laid were submitted as a chicken tax. In Gnadenfeld the tax was thirty eggs per hen.[12] By the end of 1932 not only the chicken flocks but even rabbits that had helped keep the Rahn family alive in the winter of 1931/32 were being confiscated or taxed.[13]

The harvest, usually well underway by mid-July, would normally have been expected to ease the famine, especially in view of the spring weather that seemed to portend a good crop.[14] Perhaps that anticipation accounts partly for the excessively high grain quotas to which the leadership of the local artels and the collective membership readily assented in the spring.[15] In the end the harvest fell far short of expectations and did nothing to reduce the famine.[16] In the Molotschna most of the villages, and certainly

[8] DB 1/06/1932, p. 3; MR 06/07/1932, p. 6.

[9] MR 06/07/1932, p. 6, 13/07/1932, p. 6.

[10] A frequent complaint in many of the letters from the Ukraine published at this time in MR, e.g., 21/09/1932, p. 7.

[11] DB 14/09/1932, p. 4

[12] DB 4/01/1932, p. 4. The letter is dated 5/11/1931 from Gnadenfeld and gives an especially thorough picture of conditions in the Eastern Molotschna at the end of this year.

[13] DB 21/12/1932, p. 4.

[14] MR 24/08/1932, p. 7, 21/09/1932, p. 7.

[15] DB 21/12/1932, p. 3.

[16] Colin P. Neufeldt, p. 238. In Gnadenfeld they could only meet a quarter of the expected quota. DB 4/01/1933, p. 4.

those in the eastern end of the colony, experienced a very poor harvest.[17] In Gnadenfeld, and probably in Kontiniusfeld/Sparrau as well, they were still threshing day and night in November. Although by this time in the year the villagers needed the straw to heat their houses, since nightly frosts were already occurring, it was still kept back, because a re-threshing had been decreed.[18]

The poor harvest was the result of several factors. In February there had been severe flooding as well as harsh frosts in several parts of the Ukraine that damaged or killed much of the winter wheat sown the previous autumn.[19] During the rest of the spring, particularly in the Chortitza region drought like conditions were created by temperature and strong wind.[20] The workers themselves and the horses that remained were weak and emaciated because of the starvation, so much so that even the young people and children [boys and girls] were pressed into service.[21] The quality of the seed grain left much to be desired, since in the previous fall the local authorities had ensured their own survival and the fulfilling of the procurement quotas by confiscating grain of every conceivable kind and even stealing some from the collective. The next seeding and harvest were of little concern at the end of 1931.[22] The mechanization of agriculture had also failed to reach expectations, for the tractors performed very poorly, [not surprisingly, given what HP1911 and ASR 1889 had to say about their refitting and performance in letters in 1931[23]]. As an emergency measure it was decreed in some villages that two pairs of cattle should be trained to pull wagons, an obvious admission that the MTS was not achieving theoretical expectations.[24]

[17] Lichtenau: MR 21/09/1932, p. 7; Margenau: DB 21/09/1932, p. 4; Rosenort: MR 24/08/1932, p. 7; unnamed village in Molotschna: DB 2/11/1932, p. 3; Gnadenfeld and Stulnjevo: MR 7/12/1932, p. 7, DB 4/01/1933, p. 4, Pordenau: MR 30/11/1932 p. 7.

[18] DB 4/01/1933, p. 4.

[19] DB 4/05/1932, p. 4, 12/10/1932, p. 4.

[20] Neufeldt, p. 240.

[21] DB 5/10/1932, p. 4, 2/03/1932, p. 3; MR 16/03/1932, p. 7, 6/07/1932, p. 7.

[22] DB 12/10/1932, p. 4.

[23] 03/02; 04/23.

[24] DB 21/12/1932, p. 4.

The plight of the harvesting workers was augmented by the extreme measures instituted, often arbitrarily on an *ad hoc* basis from village to village, by the local procurement commissions. The reason behind many measures lay in the grain quotas already mentioned. It seems that the Soviet leadership began such estimations, not from what was probable or even possible, but from what was needed to fulfil national industrialization and propaganda goals. During these years the Soviet Union actually exported wheat to obtain foreign currency to buy machinery, while boasting how quickly industrialization, mechanization and economic development were being achieved under their system. Such macro-economic aims, approved by the five-year plan of 1929,[25] put enormous pressure on the local communist leadership to contribute. In the autumn of 1931 or spring of 1932 they had agreed to grain production and procurement quotas in public meetings of the entire collective, where no one dared to challenge the numbers. The harvest proved such quotas to be unrealistic. For instance, in Rosenort the agreed upon grain quota was 3054 pud [approx. 50,000 kilograms] but the entire harvest did not reach 4000 pud [less than 64,000 kilograms].[26] Since most of the grain was immediately hauled away, a common occurrence through out the Ukraine in 1932,[27] there was hardly enough left over to feed the local communist officials, never mind the villagers and workers in the artels. By November food rations were only being given to actual workers in the artel; no additional food was given out for other family members, and through out much of the Ukraine local officials were demanding that the harvest be re-threshed.[28] Moreover, regular stores were devoid of all goods and simply had nothing to sell.[29] Under such conditions life would have been especially

[25] McClellan, pp. 122-130.
[26] MR 24/08/1932, p. 7.
[27] DB 2/11/1932, p. 3; 30/11/1932, p. 4.
[28] MR 16/11/1932, p. 6, 7; Neufeldt, ... p. 241.
[29] DB 2/11/1932, p. 3, 4/01/1933, p. 4.

difficult for the aged and the children. But perhaps it soon became even more difficult for the local architects of all this misery and incompetence who had originally agreed to and even pushed for the unrealistic production goals and now had to face the responsibility for the shortfalls. During the winter of 1932/3 many were purged from the local councils or the party and even had to endure the hardships that they had so callously inflicted on their fellow villagers earlier.[30]

Into this economic and agricultural morass in 1932 the Soviet leadership introduced two, what seem in hindsight, extraordinarily cynical measures. Both were aimed at obtaining further foreign currency. The first was the establishment of the Torgsins, specialty stores, probably set up originally for foreign residents, since citizens of the Soviet Union, at some point, were forbidden to shop there, and in any event could not afford it. It seems that almost everything could be bought here for gold or foreign currency.[31] Their presence in the Soviet Union was suddenly given widespread publicity.[32] The Soviet state even created a mechanism whereby foreign money could be exchanged through a Moscow bank directly to a particular Torgsin, where the designated recipients living in the USSR could buy food and other necessities for the transferred money.[33] This soon became the preferred way for people in North America to support their relatives in the Soviet Union.[34] But even this was not all it seemed. Soon there was a huge backlog of unfulfilled orders as a result of these transfers of money through the banks.[35] For many Ukrainian residents who in their weakened physical condition usually had to travel a significant distance to one of the major centres, like Zaporosche or Berdyansk[36], to find a Torgsin, a second or third trip to collect what was designated for them was very exhausting. Thus in the end the recipients were often pressured into taking items they didn't want in order to obtain the full

[30] DB 21/12/1932, p. 3. This crackdown on local party leadership was a harbinger of wide ranging purges in 1933 in many areas of the soviet apparatus.

[31] MR 14/07/1932, p. 7, 9/11/1932, p. 7, 16/11/1932, p. 6.

[32] E.g., MR 30/03/1932, p. 6.

[33] MR 24/08/1932, p. 7, 26/10/1932, p. 7, 21/12/1932, p. 6.

[34] *Ibid.*

[35] MR 6/03/1932, p. 6.

[36] It was about 75 Kilometres away and had the Torgsin nearest to Kontiniusfeld.

value for their assigned benefaction.[37]

On 10 November 1932 with great fanfare in Mennonite Publications, at least, the Soviet government announced a new emigration policy initiative. Travel passes were available for 500 rubles [about $285.00 at that time].[38] If this substantial amount of money [about five times what JP 1910 earned annually as a farm labourer at that time] was paid up front, the pass was to be issued within one to three months. If the pass was refused, the government promised to return the full pass fee. To what extent that actually happened is not clear.[39] What is clear is that such emigration policy was in conflict with another policy in place since the autumn of 1929 that anyone who tried to get permission from Moscow to emigrate was being arrested and sent to Siberia or worse. Moreover, it was also affected by Canadian immigration policy that at that time restricted entry only to wives, children under eighteen, and brides. How many Mennonites actually benefitted from this is uncertain. For the Rahn family, even if the money could have been found, Canadian legalities meant that the policy was of no benefit, since none of the individuals left in Kontiniusfeld would qualify for immigration to Canada.

Concurrent with these attempts to raise foreign cash, local Communist cells were also being pressured to raise as much foreign currency as possible. Beginning in August house searches for gold and dollars seem to have increased dramatically through out the Ukraine[40] based on information obtained via the post office of who had received money from abroad. For instance, in Gnadenfeld, if some one received banknotes or dollars directly, one needed to exchange them immediately for Soviet rubels at the post office, or spend them in the Torgsin or risk having them arbitrarily confiscated by the authorities in one of their searches.[41] In other words, it

[37] MR 19/10/1932, p. 6.

[38] MR 14/12/1932, p. 7. Repeated in MR 11/01/1933, p. 7.

[39] I have not met or heard of anyone who, or whose relatives, benefitted from this programme. Also, Colin Neufeldt assured me in an e-mail that he has not been able to verify that any Mennonites actually participated in this new immigration policy.

[40] DB 23/11/1932, p. 4, 30/11/1932, p.4, 18/01/1933, p. 4.

[41] DB 4/01/1933, p. 4.

was not safe to keep any spare cash around. For if such was found it became grounds for accusations that one was failing to support the Soviet state by hoarding what should have been freely surrendered.[42] Though the GPU participated in such searches, party officials were also keen to establish their credentials with their superiors by contributing to the goals of the Soviet state and displayed equal callousness toward their intended victims. One of the favourite torture techniques at this time was to lock the victim in a small, very hot room with many other suspects and refuse to give them air until people confessed where they had hidden their foreign currency, or what they had done with it. Even the elderly who might have a few savings in gold to help them in their last days were subjected to such tactics.[43] Not only the living were searched for heirlooms and wedding rings, but even the dead were disturbed in their graves.[44]

The first targets of such searches would, of course, have been people who were known to have connections, or received letters from abroad. Thus it should come as no surprise that no letters were received from Kontiniusfeld during this year by JP1910. Whether any were written is difficult to say since neither aunt has any recollection of that particular year. Some of the letters by other Mennonites sent from the Ukraine in June 1932, and published in the *Mennonitsche Rundschau*, seem to indicate that a guide [whether official or unofficial] as to what would be allowed by censors had been declared publicly and was suddenly of greater significance than in 1930/31.[45] In any case, the exchange of letters slowed dramatically. One writer records delight over the fact that a reply to previous correspondence had taken only eight weeks to arrive from Canada, implying that it usually took much longer. Another letter sent from the Ukraine took several months to arrive in

[42] *Ibid.*

[43] *Ibid.* Jacob A. Neufeld, *Tiefenwege,* p. 14.

[44] For a description of a house search by a party official, very much less efficient than by the GPU, see Dolot (1985), pp. 46-51. The grave robbing is described by Dolot on p. 178. It may be that part of the 1945 communist obsession with destroying completely the Mennonite cemeteries of Kontiniusfeld and Sparrau, as well as of other villages of the Molotschna, was a desire to hide the desecrations that had taken place.

[45] MR 3/07/1932, p. 6.

[46] MR 16/11/1932, p. 6, 14/12/1932, p. 6.

Canada.[46] This greater scrutiny by the censors also significantly reduced the volume of correspondence from the Soviet Union to Canada. Both the *Bote* and the *Rundschau* received only a trickle of letters for publication in 1932 compared to the volume of correspondence published in 1930 and 1931. Obviously, given the massive propaganda effort by the Soviet government to convince the world of the superiority of communist ideology and methods, not only would there be severe restrictions on what could be written, but slowing the exchange of letters provided time for the lives of people affected so dramatically for the worse to improve before word of actual conditions got out.

The effort to stamp out religious activity did not abate simply because people were starving. In fact, the famine seemed to be seized upon by some as an opportunity to attack or destroy even more of the people who held fast to some form of faith or exhibited public respect for religious festivals. As previously pointed out, the high arbitrary quotas for the submission of grain, money and meat was often specifically directed at the religious leadership. In one Molotschna village the preacher's family was given contaminated meat, which almost poisoned them. In Waldheim there were no preachers left by May; by September the church was closed because the members could not pay the 500 ruble tax required to keep it open. In Alexandertal two preachers were jailed for failing to send their children to school on a Sunday. At Ruckenau although there were still church services in October, no one preached for fear of reprisals. They sang some songs, said some prayers and read some passages of scripture. In Chortitza a teacher was purged because he failed to establish a clearly Marxist psychological profile [i.e., he did not push atheism] for each of his students.[47] Finally, in Gnadenfeld, the capital of the Eastern Molotschna, of eight original preachers in the congregation only one remained in November 1932. And he was continuously hounded by high taxes arbitrarily imposed, and then reassessed even higher amounts.[48] In the midst of these very debilitating conditions, the Rahn family survived, in

[47] MR 7/09/1932, p. 6, 25/05/1932, p. 6, 26/10/1932, p. 7, 26/10/1932, p. 7, 16/11/1932, p. 7; DB 13/07/1932, p. 6.
[48] H. Dirks, DB 4/01/1933.

that all the young people and the parents worked in the artel. This provided at least some nourishment for them. At the same time in 1934 PJ 1883 explicitly acknowledges[49] receiving aid from JP 1910 in the two preceding years. Clearly there were many massive relief efforts organized by North American and German Mennonites through out these years.[50] It is not clear which organizations JP 1910 used to get relief to his family, but I suspect that mainly he forwarded registered money through the banks to a Torgsin. He told me that he had sent them $5.00 every month as long as it was possible.[51]

Two other events also contributed to the family's survival through the winter of 1932/33. In early autumn, probably September, the oldest sister, Lena Rahn 1914, managed to obtain a post with a family in Leningrad [St. Petersburg]. These were educated people, communists, probably teachers, who wanted a German-speaking nanny to watch their children and teach them her language. The mother was German, the father Russian. Here she had enough to eat and earned 30 rubles per month. In that large city she was able to buy necessities, not available in the agricultural villages, and send them on to Kontiniusfeld. The second change to the family came with the departure of the father, PJ 1883, late in 1932 for the Caucasus. He had been under constant scrutiny by local Communists [and probably the GPU] since early 1930. Some of this attention originally resulted from his possessing Dollars. Later in spring 1931 he was indicted for his failure to meet the required grain contributions. It seems that he left very quietly and hid himself with his brother David Rahn 1878. His nephew David David Rahn 1924 [now deceased] told my aunt Anni Rahn Wiens 1927 that he

[49] See letter 1934/03/29.

[50] During 1932/33 the *Mennonitsche Rundschau* published long lists of donors and the amount donated to the Mennonite relief efforts at least once or twice per month. For an full account of these efforts, often spearheaded by the author himself, see B. H. Unruh, *Fuegung und Fuehrung im Mennonitischen Welthilfswerk*, 1920-1933. And of course, there was also the work of Jakob Kroeker, referred to earlier. See commentary on letter 1931/04/10.

[51] Aunt Justa (Rahn) Prei 1922 in her memoirs, *Mein Begleiter Jesus Christus*, p. 12 says that JP 1910 generally sent them ten dollars every second month, though it did not consistently get through to them every time.

remembers PJ 1883 hiding at their place in clothes closets, the haymow and the root cellar while eating frozen kernels of corn which he picked from cobs gleaned from the collective farm. The scrutiny in the Caucasus was less intense, the grain procurement commissions not as rabid and, since he was unknown there, the eye of the GPU was not as alert. Since he was deemed the black sheep in Kontiniusfeld [probably in no small part because he continued to participate in church services and bible studies as well] and given what was happening to other families whose men were connected to the church, by disappearing from the village he took some of the pressure off his family. At the same time he did not deprive any of the others of what little food they earned or grew.

And so 1932 came to an end with Lena 1914 working in Leningrad, PJ 1883 hiding in the Caucasus, Heinrich 1911 working in the MTS, Hans 1913 working in the collective cow barn, Aganetha Schartner Rahn 1889 working in the artel as demanded as well as husbanding what little they had been able to keep of their own vegetables, rabbits and domestic fowl. Justa 1922 and Mika 1924 received some meals at school. Thus only Anni 1927 had no source of food because she was too young. All survived the extreme famine in the winter 1932/33.

1933–35
in Kontiniusfeld

Family name abbreviations used in discussion of the letters

ASR 1889	Aganetha Schartner Rahn 1889 – 1968
PJ1883	Peter Jacob Rahn 1883 – 1942
JP1910	Jacob Peter Rahn 1910 – 1991
HP 1911	Henry Peter Rahn 1911 – 1996
JohP 1913	Johann (Hans) Peter Rahn 1913 – 1944
HelP 1914	Helena Peter (Rahn) Pauls b. 1914
JusP 1922	Justa Peter (Rahn) Prei Geddart 1922 -2004
Anni P 1927	Anna Peter (Rahn) Wiens b. 1927

1933-1935 in Kontiniusfeld

Dez. 23.1933[i]

Wird es auf Erden ein Wiedersehen geben?

Lieber Bruder, Ich will ein paar Zeilen schreiben an Dich. Justel und Mika gehen für Änni durch, denn sie wird böse. Sonnabend habe ich eine Karte bekommen, dass die nicht Recht haben. Ich wollte mit Kaninchen tauschen bei J. S. Dann er wollte noch Rubel und 25 Kopeken haben und das tu ich nicht. Der kleine Aron hat auch Kaninchen und davon ist ein Krüppel. Schick mir auch ein paar Blätter. Johann.

Es[ii] war? Ich habe zwei? holen? und zusammen gesammelt und dafür habe ich 10 bekommen und für fünf Kopeken habe ich Justel einen Bleistift gekauft zu Weihnachten. Ich fahre wieder Morgen auf Cursen. Habe auch?? Kamm immer auch zum Mittage und Nacht Hause u.[nd] acht Tage musste ich arbeiten. Bis jetzt bin ich noch immer zum Sonntag nach Hause gekommen. Jetzt soll ich zwei Sonntage dort bleiben u.[nd] den

[i] Entweder November oder Dezember. Das ist auf dem Brief nicht klar geschrieben, aber auf Seite zwei erwähnt der zweite Brief von HP 1911 Weihnachten und eine Kurse bis zum ersten März.

[ii] Zeite zwei.

1933 12 23[1]

Will we meet again here on earth?

Dear Brother, I want to write a few lines to you. Justa and Mika are at odds with Anni, because she gets angry. Saturday I received a card that contains an error.[2] I wanted to trade rabbits with J. S.[3] but in addition he wanted a Rubel and 25 Kopeks, but I won't do that. The little Aron also has rabbits, and one of those is a cripple. Send me also a few sheets. Johann [JohP 1913]

It was ? I had two? to get? And gathered together? And for those I received 10, and for five Kopeks I bought Justa a pencil for Christmas. Tomorrow I am going off on courses again. Also have ??[4] Always came home at noon and for the night. For eight days I had to work. Until now I have always been allowed to come home on Sunday, but now I am supposed to remain there for two

[1] These are snippets that have been translated. The month of writing is uncertain but the year and the day are on the letter. The numerical date was probably added later. The numbers were tacked onto and partially written over the letters De. The next letter is probably a "z" for *Dezember* in abbreviated form or an "n" for *Den*.

[2] I suspect that this was a card concerning his induction into the army referred to by other relatives in later correspondence.

[3] Johann or Jacob Siebert according to my aunts.

[4] Probably something to do with military service.

dritten komm ich nach hause. Werde dann sehen wie es geht. Soll bis zum ersten März lernen. Gruß, dein Bruder Heinrich Hast[iii] Du noch Hoffnung, dass es noch auf Erden wird ein Wiedersehen geben? Ach könnten wir gerettet werden um der unschuldigen Kinder halber. Aber wer weiß, was uns droht? Dunkel schaut einer in die Zukunft. Wer weiß, was uns erwartet. Aber wir stehen ja in Gottes Hand. Es wird ja kein Haar von unserem Haupt fallen ohne seinen Willen. Bis heute haben wir noch nicht zu klagen, wenn wir auf viele schauen, denen es noch schlechter geht. Zum Schluss sei herzlich gegrüßt von den Deinen. Grüße die Großeltern. Ohm hat von David Dirk. einen Brief erhalten. [PJ 1883]

[iii] Auf ein anderes kleines Papier. JP 1910 hat diese zwei kleine Stücke Papier zusammen behalten und den Datum darauf geschrieben.

Sundays and then come home on the third. Will see how it turns out. Am supposed to study until the first of March. Greetings, your brother, Heinrich. [HP 1911]

Do[5] you still have hope that we will meet again here on earth? If only we could be saved for the sake of the innocent children [Justa, Mika and Anni]. But who knows what is threatening us. One peers darkly into the future. Who knows what awaits us. But, after all we remain in God's Hand. Not a hair of our head shall be lost against his will. Up to today we have nothing to complain about when we look upon many others who experience worse things [than we]. To close, hearty greetings to you from yours [family, relatives]. Greet the Grandparents. Ohm [Hein] has received a letter from David Dirk[sen]. [PJ 1883]

[5] This is a separate fragment that has only one side filled in. There may have been more to this letter. It is unusual that only one side is filled in. They usually utilized all the paper because it was in such short supply. The handwriting is that of PJ 1883. The request to greet the Grandparents [BB Janz 1852-1932] may mean that this fragment should have been placed with some much earlier letter since by 1933 BB Janz 1852 had died.

1934 02 27

Nach langem Warten endlich einen Brief

Kontiniusfeld

Lieber Jakob, da wir Heute nach langem Warten endlich einen Brief von Dir erhalten haben und zwar den vom 28.1., so will ich versuchen, ihn auch gleich zu beantworten. Wir haben schon sehr darum gebetet, um mal wieder was von Dir zu hören, aber auch gebetet, dass Du von unseren Briefen etwas erhalten möchtest. Lag es mir doch auf dem Herzen, dass Du Dich sehnen würdest, weil keine Nachricht von uns kommt und Du so allein, so weit von uns getrennt bist, obzwar Heinrich und Lena auch nicht zu Hause sind. Letztere in Leningrad, ersterer in Kie.[v] in der roten Armee bei der Artillerie. Hans musste auch schon 14 Tage lernen, er glaubt, es wird ihm gut gehen, dienen. Justa, Mika und Änni gehen zur Schule, letztere nur Nachmittags in der Dorfschule, so dass es bei uns schon leer wird im Haus. Es ist Nacht und ich wache im Lagerraum. Bin nun schon bald ein Jahr Wächter und nun vor Langeweile schreibe ich an Dich. Mama

1934 02 27

After a Lengthy Wait, Finally a Letter

Kontiniusfeld

Dear Jacob, Since, after a lengthy wait, we finally received a letter from you, and that from the 28/01, I shall undertake to answer it right away. We have prayed much that we might hear from you once again, but also that you might receive something of the letters [we have sent]. It was already heavy upon my heart, how homesick you would be, because no news was arriving from us, and you, separated from us, so far away; though Heinrich and Lena are also not at home. The latter is in Leningrad, the former in Kie.[v] in the Red Army as an artilleryman. Hans also had to take 14 days training. He thinks it will be enjoyable to serve.[6] Justa, Mika and Annie are going to school, the latter only in the afternoon to the village school (*Dorfschule*). So it is becoming empty at our house. It is nighttime and I am keeping watch in the storerooms. I have already been a watchman for almost a year, and now I am writing to you during a time of boredom. Otherwise Mama

[6] According to my aunt, Justa (Rahn)Prei Geddart 1922 [JusP 1922, *Mein Wegbegleiter Jesus Christus,* p. 14] JohP 1913 was sent to Vladivostok to serve there in the army in 1935. After his two-year service he remained in the East to work as a farm labourer. During the WW II one of the Rahn relatives banished to an area near the Amur met him in an internment camp for German ethnics in Khabarovsk where he died of starvation. In light of his later fate this present anticipation is most ironic.

wollte sonst schreiben, dann verläuft so schneller die Zeit. Wir sind, Gott Lob und Dank, schön gesund, Lena auch. Nur von Heinrich haben wir schon fast einen Monat keine Nachricht, so dass wir gar nicht wissen, wie es um ihn steht. Ohm Heinrich ist im Waldheimer Krankenhaus, hat Karbunkel, wird wohl sein letztes sein. War schon sehr verkommen, voll Ungeziefer, ist in letzter Zeit bei uns nicht gekommen. Er wurde im vorigen Jahr bestohlen, als Heinrich bei ihm zur Nacht war und Heinrich gerade die Diebe verstörte, sonst wäre es noch ärger geworden. Heinrich kam dann des Nachts nach Hause, war ganz verängstigt, musste aber gleich wieder mit Hans hingehen, damit sie Ohm nicht noch totschlagen sollten und jetzt sagte er, Heinrich sei auch darunter gewesen unter den Dieben. Dann kam er mal her und dann zankten sie sich und seit der Zeit ist er nicht mehr gekommen. Heinrich war dann in Todesgefahr, er lag gerade auf

already wanted to write, but now [in writing] the time passes more quickly. We are, praise and thank God, all healthy, even Lena. Only from Heinrich we have heard no news for almost a month. So we have no idea how it is going with him.[7] Ohm Heinrich [Janz] is in the Waldheim hospital; has carbuncles[8]; will probably be his last [ailment]. Was already very deteriorated, full of vermin [lice, etc.], was no longer coming to [visit] us in the recent past. Last year he was robbed, when Heinrich was staying the night with him and disturbed the thieves. Otherwise it would have been even worse for him. Heinrich came home at night, was completely overcome with panic, but had to return immediately with Hans so that the thieves would not beat Ohm to death. Now he [Ohm Hein] said that Heinrich had also been among the thieves. After that he came here again; they quarrelled. Since then he has not come back. At that time Heinrich had been in serious danger of losing his life, for he was sleeping

[7] According to my aunts, HP 1911 caught typhoid fever, was released from military service and returned to Kontiniusfeld, where he may have driven tractor again for the MTS, but more likely went to work in the collectivized cow barn.

[8] A carbuncle is an abscess larger than a boil, usually with one or more openings draining pus onto the skin. It is usually caused by bacterial infection, most commonly *Staphylococcus aureus*. The infection is contagious and may spread to other areas of the body or other people. According to my aunts, Ohm Heinrich did not die, but disappeared from Kontiniusfeld and died later in Memrik.

der Kiste, wo das meiste Mehl drinnen war. Und Eisen und Beil lag da herum, wie leicht hätte man ihn erschlagen können. Er sagte auch jeden Abend, wenn er hingehen wollte, wollen noch erst beten. So Angst war ihm, aber er ging dann doch. Und nun belog er ihn so im Dorf. Aber Undank ist der Welten Lohn. Matth. 5.11. Wir haben Brot und werden auch ausreichen, wenn es uns bewahrt bleibt. Grüße die Großmama und die Onkels und Tanten, so auch Dirksens, Koppen, Peters. Letztere ihre Geschwister sind gesund. Solltest Du mal Franz Braun treffen, dann grüße auch den von mir. Werde Schluss machen. Verbleiben Deine für Dich betenden Eltern und Geschwister. Aufwiedersehen. Wenn nicht hier, dann droben bei Jesu. Die letzten 10 Dollar haben wir erhalten fürs Auto. Lena erhält 40 Rubel den Monat. [PJ 1883]

on top of the chest in which most of the flour was stored. Pieces of iron and a hatchet lay around there, how easy it would have been to beat him to death. Every evening when he [HP 1911] was about to go there [to Ohm Hein], he always said, "Let's pray first." That was how afraid he was, but still he went anyway. And now he [Ohm Hein] lied about him thus in the village. But thanklessness is characteristic of the world's wages. Matthew 5.11.[9] We have bread that will probably last [through the winter] if it can be kept safe. Greet Grandma[10] and the aunts and uncles, as well as the Dirksens, Koops, and Peters.[11] The relatives of the latter are well. Should you ever meet Franz Braun, then greet him for me. Will come to an end. Remain your parents and relatives who pray for you. Till we meet (*Aufwiedersehen*), if not here then above with Jesus. We received the last 10 dollars for the auto.[12] Lena receives 40 rubels per month. [PJ 1883]

[9] *Blessed are you when men revile you and persecute you and utter all kinds of evil against you falsely on my account.*

[10] Mrs. B. B. Janz, [Helena Vogt Penner 1854].

[11] These Peters are the parents of our aunt, Tina Wiebe, married to our mother's brother, Henry Peter Wiebe.

[12] This is very peculiar, for they had no automobile, not even to be hired. This must be an abbreviation of some sort, perhaps for *automat*, meaning some form of money transfer.

COMMENTARY

The fact that the writer needs to explain what he has been doing for the past year, as well as his reference to what transpired last year between Ohm Hein and HP 1911, reinforces that there really has been little or no correspondence between Canada and Kontiniusfeld for some time. Obviously, the restrictions on mail delivery occurred in Russia, for it is clear that the family in Kontiniusfeld had written some letters, but none seem to have arrived, for there is a long gap in JP 1910's preserved letters from the family. Moreover, in JP 1910's memoirs, in the paraphrases of the letters he received there is no mention of any letter after 22 November 1931 until 27 February 1934 except for the fragments and short notes from December 1933. Thus it seems that only near the end of 1933 did the correspondence begin to move again. That may explain the rather fragmentary and odd nature of the letters and notes sent in December 1933. They were probably a test to see if anything was getting through to Canada.

The reference to receiving the money reinforces what my father [JP 1910] told me about sending money regularly to his relatives in the early 1930s. However, whether this was sent directly or through a relief organization such as Jakob Kroeker's *Licht im Osten* is not clear. My aunt, Justa [JusP 1922], in her memoirs[13] states explicitly that the family received ten Canadian dollars every two months, and that at a later date they were required by the GPU to send the money back. There is a wide range of possibilities for the money transfer. The main one of course was through the State Bank in Moscow that then authorized a particular Torgsin to sell goods to a person and sent a voucher for a certain amount in a particular currency to the person in question. In addition to the problems with availability of foodstuff in the Torgsin noted earlier, the vouchers often took three months or more to arrive or, with increasing frequency in 1934, did not arrive at all. Sometimes the vouchers were withheld until taxes that were in arrears were paid or the recipients were pressured to refuse the vouchers as well as any packages that might be sent to them. Thus some people in the Ukraine preferred to receive aid directly in cash [but not in German currency] in the envelope along with

[12] *Mein Wegbegleiter Jesus Christus,* p. 12.

the letters. The cash, provided the letter wasn't opened, allowed them to avoid the attention of the GPU, and use the money to buy things privately.[14] I think that some form of cash was the most likely way that money came from Canada to the relatives in Kontiniusfeld. They lived in the Eastern part of the Molotschna far from both Melitopol and Berdjansk where the nearest Torgsins were located. As noted by other letter writers from the same region,[15] to get there would require a train trip of several hours. Often after arrival late in the morning or early afternoon there would have been little for our relatives to buy since those living nearby would already have purchased most of the food supplies available in the stores.[16] Moreover, some people who had purchased flour at a Torgsin and were intending to carry it home were refused train transportation on the grounds that to transport flour by train was illegal.[17] In considering what my father told me about sending five dollars monthly and what my Aunt Justa writes in her memoirs, I think it likely that most of this money was sent in cash, for this would have given them the greatest flexibility in spending it.

The reference to JP 1910's Grandma also confirms that some letters from Canada were getting through somehow. BB Janz 1852 had already passed away on 8 January 1932. Since PJ 1883 knew about that, some letters from Canada must have reached Kontiniusfeld, even if none seem to have arrived in Canada from there for most of the previous two years.

Whether this interruption in communication was due entirely to Soviet intervention is questionable. For some period of time in the early 1930s PJ 1883 had disappeared from Kontiniusfeld, partly because party authorities were constantly harassing him. The other members of the family seemed to have earned their regular rations from the Artel. Since he still refused to join the Kolkhoz, he always seemed to be in trouble [probably because of his commitment to his faith as well] and so jeopardized the family's wellbeing. After he had left, one night the GPU and local party members surrounded the family home in Kontiniusfeld and searched it thoroughly, even scattering

[14] DB 23/03/1934, p. 4, 31/10/1934, p. 3, 7/11/1934, p. 4; MR 17/10/1934, p. 7.

[15] DB 19/09/1934, p. 4.

[16] *Letters from Susan,* p. 124.

[17] DB 23/03/1934, p. 3.

the hay in the loft and bringing in searchlights to inspect the well.[18] By going to the Caucasus and hiding with his brother, David Jacob Rahn b. 1878, PJ 1883 removed the need to feed him from the family rations. His nephew, David David Rahn 1924 [now deceased] told my aunt, Anni Rahn Wiens 1927 that he remembered PJ 1883 hiding at their place and eating frozen kernels of corn gleaned from the fields. Aside from the fact that his brother lived there, another reason for going to the Caucasus was that the program to collectivize agriculture was much more rigorously applied in fertile areas like the Ukraine and less advanced in the Caucasus. Thus the authorities would perhaps have been less omnipresent there where he was unknown. Apparently he earned some money there that he somehow sent home to the family.[19]

While we know little about how the family survived the winter of 1931/32, it is clear that PJ 1883 was still with them at that time. My aunt, Lena Rahn Pauls 1914, left Kontiniusfeld for Leningrad in September 1932 to work as a household domestic. At that point she remembers her father, PJ 1883, was still in Kontiniusfeld and for a time opposed her going. Eventually she left. Shortly after that he journeyed to the Caucasus and spent the winter of 1932/33 there.[20] PJ 1883's departure from Kontiniusfeld along with the ever-present famine may have been other reasons why fewer or no letters were written in 1932/33. No one wanted to draw any attention to the family whereabouts, and most certainly, PJ 1883 was not going to give himself away, even if he had access to pen and paper.

Based on the comments in the letter above he must have returned to Kontiniusfeld in the late spring of 1933. It seems that the pressure to collectivize had also increased in the Caucasus. So he returned to Kontiniusfeld, registered with the GPU and officially surrendered his land to the collective, which he had long opposed.[21]

[18] JusP 1922, *Mein Wegbegleiter Jesus Christus,* p. 13.

[19] *Ibid.*, p. 11.

[20] My aunts, Lena Rahn Pauls 1914 and Anni Rahn Wiens 1927, do not remember clearly when he left and when he returned; Lena, because she was gone from Kontiniusfeld and didn't return there for several years and Anni, because she was too young.

[21] JusP 1922, *Mein Wegbegleiter Jesus Christus,* p. 13.

1934 03 29

Wärst Du nicht da gewesen...

Lieber Jakob, Da wir schon wieder geraume Zeit keine Nachricht erhalten haben, so will ich dann wieder versuchen Dir ein Lebenszeichen zu senden. Durch Gottes Gnade und Güte erfreuen wir uns einer schönen Gesundheit, welche wir Dir auch von ganzem Herzen wünschen. Dein Brief, welchen wir erhielten, war vom 28. Januar. Freuten uns, endlich mal von Dir etwas zu lesen. Wie war es doch so schön, als wir alle Wochen etwas von Dir lesen konnten. Und jetzt ist es so, als ob nicht mehr möglich ist, etwas hinzuschreiben, und ob auch nichts mehr her darf kommen. Nun möchte der liebe Gott Dir dieses Schreiben in Deine Hände führen, ist mein – oder richtiger – unser Wunsch. Dein ältester Bruder ist in der Armee, der jüngste kommt nun zwei Jahre dran. Lena ist nicht mehr zu Hause und Justa das nächste Jahr auch nicht mehr, wird sie doch schon das kommende Jahr in Gnadenfeld zur Schule gehen. So ist bald die ganze Familie verstreut. Und ob wir uns noch

1934 03 29[22]

Had You not been over There…

Dear Jakob, Since we have received no news for a significant time, I will try to send you a token that we are still alive. Through God's goodness and grace we enjoy good health, something we also heartily wish for you. Your letter that we received was from the 28th of January. Were happy to finally hear something from you. How pleasant it was when we were able weekly to read something from you. And now it is as if it is no longer possible to write anything to you, nor is it allowed that anything be sent here anymore.[23] Now may our loving God guide this written material into your hands is my or, more correctly, our wish.[24] The older of your brothers, Henry is in the armed forces, the military; the youngest brother [John] will be liable for service in two years. Lena is no longer at home [in Leningrad]. And Justa next year will also not be [home], for in the coming year she will have to go to school in Gnadenfeld. And so the whole family will soon be scattered.

[22] In Aunt Anni's autobiography, *Gepriesen sei die Gnade Gottes*, she reproduces four letters [1930 05 18; 1930 06 22; 1934 03 29; 1934 04 28]. I believe that my father, JP 1910, gave her this one. I sent her copies of most of the others. Between her and H. Woelk these letters all were transliterated which saved me much work.

[23] This comment reaffirms that censorship and interference with foreign correspondence has increased.

[24] Obviously, this prayer was answered.

mal wiedersehen werden, ist eine große Frage, ich meine in diesem Leben. Nun sollte der Gott der Liebe es so beschlossen haben, uns das vorzuenthalten, wenn wir uns dann aber dort vor seinem Thron wiedersehen möchten, das doch das Bessere ist. Wenn wir eben können, dann werden wir uns fotografieren lassen und Dir unsere Fotografie schicken. Aber würden uns auch gerne Deine wünschen oder wir wünschen sie uns, damit wir Dich wenigstens mal sehen könnten, so wie es Dir jetzt sieht. Habe mich in letzter Zeit schon so oft gesehnt, Dich zu sehen und zu sprechen, aber vergebens. Und Dir geht's doch wohl so? Mir scheint's so. Es hat mir einfach gejammert über Dich, dass Du so allein musst sein. Ein Glück, Du hast dort Großmama und Onkel, Tanten und Bekannte, aber es sind nicht Eltern und Geschwister. Ist es nicht wahr? Oder geht es Dir nicht so? Und doch danke ich Gott, dass Du da bist, habe es schon manchmal bereut, dass Ihr nicht alle vier da seit, oder wir alle. Nun Gott

Whether we will all see one another again is a big question—I mean in this life! How we would rejoice. However should the God of Love have decided to postpone this [family reunion], then when we will be reunited before his throne that would still be the best course. If we are at all able, we will have a family photograph made and sent to you. But we would also gladly have one of you. And we desire one of you so that we could at least see you for once, what you look like now. In the recent past I have often longed to talk with you or to see you, but in vain. And for you all is going well, thus? It seems that way to me. I have recently simply felt so much pity for you that you have to be so alone! Luckily you have Grandmama and uncles and aunts[25] and other acquaintances. But they are not parents or siblings. Isn't that right? Or don't you have that experience? And still I thank God that you are there; have often regretted that not all four of you[26] are there,

[25] The relatives referred to are Mrs. B. B. Janz [Helena Vogt Penner 1854], B.B. Janz 1877, J. B. Janz 1884, Mrs. A. Toews [Helena Janz 1894] and their spouses who all lived in the Coaldale, AB area.

[26] Jakob, Henry, John, and Lena, the four children from the first marriage. At first glance this might appear to make a distinction between members of the blended family. But that is not the intent here. This is the only place where such a possible attitude appears. Elsewhere in the correspondence there is never any question that they are all one family. Rather the reason for singling them out here is that they have all left home and are not part of the household any more and, hence, PJ 1883 is worried about their well-being.

weiß, wozu es gut ist, dass es so ist wie es eben ist. Wärst Du nicht da gewesen, dann wären wir jedenfalls schon nicht unter den Lebenden, aber Gott wusste, was er tat, als er Dich von hier weg führte. Ich denke dann oft an Joseph, der musste auch weg wegen den seinigen. Ei, wie geht es Dir? Hast Du Arbeit und wo? Wir haben alle Hände voll zu tun. So dass es einem schon so über wird, alle Tage arbeiten und arbeiten. Wenn ich auch nicht so physisch arbeiten brauche, aber mal alle Nacht wachen, dann wird einem doch so müde und so matt, dass manchmal die Seele kein Leben mehr hat. Sagt das duster, aber es ist auch so, die Nerven schmerzen so stark, dass ich gar nicht schlafen kann. Dann wenn ich schlafen will, muss am Tage alles still sein im Hause, so dass sich niemand im Hause hören lässt, sonst wird von Schlafen

or all of us! Now God knows to what purpose things are as they are. Had you not been over there [in Canada], then probably we would no longer be in the land of the living.[27] But God knew what He was up to, when he led you away from here. I often think of Joseph, who also had to leave, for the benefit of his [relatives]. So how is it going for you? Do you have employment? Where? We have our hands full with work. So much so that one is weary of the daily grind of work and more work. Although I do not need to do physical labour, yet always to be awake all night, then one becomes so tired and so listless, that sometimes the soul has no more life; call it gloomy. Often I have such strong nerve pain [nervousness, shingles?] that I can't sleep at all. Then when I want to sleep during the day, everything must be absolutely still in the house, so that no one is heard. Otherwise

[27] This reinforces what my father, JP 1910, often mentioned about sending money monthly to his family in Kontiniusfeld. It also supports and expands the statement of Colin Neufeldt in his thesis [p. 111f.] that food packages and money-containing letters *"saved their* [exiled Mennonites] *lives."* Not only those in exile in Siberia and elsewhere, but also those who remained behind in the villages of the Ukraine required support, since virtually everything, even what they produced for themselves after working full time in the artel, was taken from them. Often only money would reduce the pressures of the oppressive Stalinist five-year plan to demonstrate to the world how productive the communist system was.

gar nichts. Und dann sehne ich mich, schon aufgelöst zu sein, dass ich oft so denke, wie der Dichter sagt. "Wann schlägt die Stunde, ach wann darf ich gehen, Heim ach nur Heim." Aber wegen der lieben Kleinen will man dann doch hier bleiben, bis auch sie groß sind. Habe schon oft gesagt, ach könntet Ihr alle selig sterben, ich wollte auch schon der letzte sein! Nun, wie es unserem Herrn Jesu gefällt. Werde schließen mit Schreiben und den Brief Morgen auf die Post geben. So sei denn noch herzlich gegrüßt und im Geiste geküsst von deinen für dich betende Eltern und Geschwistern. Auf Wiedersehen! Markus 13.23-37. [PJ 1883]

there is no sleeping [for me].[28] And then I already long to be freed [from this]. Thus I often think as the poet writes, "When comes the hour, alas, when may I depart home, only home." But because of the beloved little ones [Justa, Mika, Anni] one still wants to remain until they too have grown up. Have already often said, if only all of you could die in blessed salvation (*selig sterben*) I should readily be the last [to pass away]. Now, however, as it pleases our Lord Jesus. Will quit writing and take the letter to the post tomorrow. So then hearty greetings to you and a kiss in spirit from your parents (who pray for you) and your siblings. Aufwiedersehen! Mark 13.23-37.[29] [PJ 1883]

[28] According to my aunts PJ 1883 often had difficulty sleeping at night and would frequently walk through the village and sing hymns or other songs at the windows of fellow villagers' homes.

[29] *... And they will see the Son of Man coming in clouds with great power and glory. ...Watch, therefore...*

COMMENTARY

The comments at the beginning of this and the subsequent letter support what had already been guessed from the previous letter, namely, that there had been significant interruption in the mail service, especially with regard to letters sent from the Ukraine. PJ 1883's remark that without his son's presence in Canada they would not have survived indicates that at least some of JP 1910's letters with money [however it was delivered] were getting through. Nevertheless even that seems to have been sporadic. The remark about how happy they are to *finally hear something* from him taken with the later comment about *still being in the land of the living* because he [JP 1910] was *over there* imply that in 1933, and probably earlier, the family had not heard from him regularly but clearly had received significant support from him. In the next letter [1934 04 28] PJ 1883 quotes JP 1910 as saying in a letter dated 28 January [1934] that he had not heard from them for more than seven months. Thus the last letter or note PJ 1910 had received was from about June 1933. That letter is not among our father's papers.

Once again, as in the previous and subsequent letter, the writer finds it necessary to reiterate what he had already reported less specifically in the previous letter about what has transpired with each member of the family. Perhaps most surprising is the statement that Lena [Lena Rahn Pauls 1914] is in Leningrad. This move had occurred in 1932, at least a year and a half before the present letter was written and gives us some idea of how long it had actually been since there was a regular exchange of letters between Kontiniusfeld and Coaldale. If this had been reported in earlier letters it would not have been necessary to emphasize that she was *no longer* at home. Moreover, while PJ 1883 was in hiding, it is likely that little or nothing was being written, especially since the last letters from 1932 already complain of a lack of letter writing materials. Finally, as already noted earlier, at that moment the family certainly did not want to attract any attention. Letters to or from a foreign country would have done that.[30] Hence the exchange of letters was curtailed for personal reasons as well as government policy.

[30] E.g., see MR 17/10/1934, p. 7, where the writer asks that all future correspondence be sent to a relative in a different village.

Clearly, the communist authorities increasingly viewed the exchange of letters with people outside the Soviet Union as an act of sabotage, particularly in 1932 as the failed harvest and harsh quotas for grain delivery brought on a very severe famine. People who corresponded with relatives or friends in other countries and complained about the famine were often designated as "enemies of the state" for this "Kulak" need for contact with residents of non-proletarian states. By restricting letter exchanges in 1932/33 the Stalinist regime was trying to prevent word of the deplorable famine from reaching the outside world, for Stalin wanted to continue to sell grain to the rest of the world in order to ensure a continuous supply of foreign currency. Clearly the restrictions on foreign correspondence curtailed the letters received by JP 1910 as well as other Mennonites in Canada. It is noteworthy that perusal of the *Mennonitische Rundschau* and *Der Bote* for the years of 1932/33 shows a significant reduction in the publication of letters received from the Soviet Union, in comparison with the number published in 1930 and 1931. The restrictions that led to this reduction became even more severe in 1934.[31]

The despondency in this letter reflecting the disappearance of hope of emigration is further augmented by the obvious pressure that the writer feels not only from the physical demands, but also from the sense of being constricted by the system. The depression he feels is related to the already mentioned scattering of his family and the family's lack of control over their future. So burdensome had this become that it was finding expression in physical disease [perhaps, shingles]. But in this rather dark hour he turned, as always, to what had been his greatest comfort, as reflected in many of his previous letters, poetry and his faith.

[31] MR 28/02/1934, p.14, 17/10/1934, p.7

1934 04 28

In diesem Jahr haben wir noch nicht gehungert

Lieber Jakob! Da wir schon wieder eine geraume Zeit keine Nachricht von dir haben, so will ich versuchen dir wieder ein Lebenszeichen zu senden. Wir sind, Gott sei Lob und Dank schön gesund und wünschen dass dir dieses Schreiben bei guter Gesundheit antreffen möge. Der Herr wolle dir selber diesen Brief in die Hände liefern, denn in deinem letzten Brief schreibst du dass du schon über 7 Monate keine Nachricht von uns hast. Den Brief hattest du von Grosmama aus geschrieben, wie du auf Besuch warst. Sie selbst hatte auch noch etliche Zeilen beigelegt. Schade dass das Verkehr gestört ist. Ging es doch so schön alle Woche einen Brief lesen, und nun kann man Monate warten bis man einen erhält. Ist es deshalb das du eine Zeit lang eine andere Adresse hattest oder ob man die Briefe nicht mehr durch lässt. Ich kann mir

1934 04 28[32]

During this Year we have not yet been Starving.

Dear Jakob! Since for a long time once again we have had no news from you, I shall try once again to send you a token of our being alive (*Lebenszeichen*). Praise and thank God we are in good health and hope that the arrival of this correspondence will also find you in good health. May the Lord himself deliver this letter directly into your hands, for in your last letter you write that for more than 7 months you have received no news from us.[33] You had written that letter from Grandmama's[34] place, when you were there on a visit. She herself had also added a few lines. Sad that communication has been interrupted.[35] Wasn't it pleasant to read a letter every week, and now one can wait for months before one arrives. Is it because for an extended period of time now you have had a different address, or is it that the letters are not allowed through any more? I am unable

[32] Original letter found in Anni Rahn Wiens 1927's autobiography.

[33] In the letter 1934 03 29, PJ 1883 mentions receiving a letter from JP 1910 dated Januuary 28th.

[34] Mrs. Helena Vogt/Penner Janz 1854, wife of Benjamin B. Janz 1852.

[35] This explains why there is such a large gap between the last letter received from Konstiniusfeld in 1931 and the next letter received in 1933. After that, of course, the mail was never as regularly exchanged again as per the next sentence.

das nicht erklären. In der vergangnen Nacht schrieb ich an Heinrich. Der ist ja im Dienst in der Roten Armee. Hans besorgt wieder Kühe. Lena ist noch auf der alten Stelle, es geht ihr gut, hat nichts zu klagen. Erhält 40 Rubel im Monat. Schickte den Kindern und Hans etwas zu Weihnachten, auch Mama ein paar Strümpfe, Mich 5 Rubel, die nahm ich und bezahlte Heinrichs Paket was wir ihm sandten. Ich bin Wächter, darf dann für mich Gottes Wort lesen. Lieder lernen auswendig, so das mir die dann Keiner nehmen kann, wenn ich sie erst auswendig kann. Denn in 14 Stunden ist einem schon nach Zeit vertrieb und so vergeht die Langweile. Und was schaffst du? Du Fragst im letzten Brief noch nach die zehn Dollars ob wir sie nicht erhalten haben. Die haben wir erhalten.

to explain this [irregularity] to myself. Last night I wrote to Henry who is, of course, serving in the Red army. Hans [JohP 1914] is again tending the cows.[36] Lena is still in her former place.[37] She is faring well, has nothing to complain about. Receives 40 Rubels per month. Sent the children and John something for Christmas, also sent Mama a pair of stockings. Me, 5 rubels, that I used to pay for a package that we sent to Henry. I am a watchman[38] and so have opportunity to read God's word, learn songs by heart so that no one can take them from me once I know them from memory. For during 14 hours one seeks for something to while away the time and [doing] this dispels the boredom. And what are you working at? You asked in your last letter about those ten dollars, whether we had received them. We did. But during

[36] JusP 1922, *Mein Wegbegleiter Jesus Christus,* p. 12 records how JohP 1913 sometimes brought home the biscuits made of sunflower seeds meant as feed for the cows.

[37] Regular employment in Leningrad since September 1932.

[38] It is interesting to note that although PJ 1883 had resisted entering the Kolkhoz until 1933, nevertheless he was considered trustworthy enough to serve as watchman over the storerooms. Clearly the communist authorities recognized honesty when it was in their interests. I asked the aunts whether he never brought food home for the family since he was watching over it. They said that he had discussed this matter with another fellow church member and they had decided that even in such circumstances one must fulfil one's duty to the state honestly.

Aber in diesem Jahr haben wir noch nicht gehungert. Und wenn wirs halten können, werden wir auch nicht hungern. Nur mit der Kleidung ist es bei uns im Argen. Anni muss auch schon in die Schule gehen aber hat nicht was anzuziehen, ein Kleidchen und das ist schon ganz beflickt, so das sie Sonntag für Sonntag zu hause sitzen muss.

Aber Du darfst Uns Nichts schicken. Vor einigen Tage hatte Anni deine Fotografie in der Hand und dann sah ich nur wie sie drückte und küsste. Wir alle sehnen uns dich wieder zu sehen. Nun Gott weiß es ob es noch mal werden wird? Ist nicht ausgeschlossen, da doch mit USA normale Beziehungen sind. Wie geht es dir? Was schaffst du? Hast du Arbeit? Oder musst schon hungern? Wir haben schon 10 Hühner. Auch ein Ferkel haben wir schon, wenn es am leben bleibt dann werden wir im nächsten Herbst auch was schlachten. Schließe; Nebst Gruß deine für dich Betende. [PJ 1883]

this year we have not yet been starving. And if we can keep it [food, crops, produce] we won't be [in the future]. Only with regard to the clothing do we find ourselves in need. Anni also has to go to school now but has nothing to wear; one dress that is patched, in so many spots, that Sunday after Sunday she has to stay at home.[39] But you must not send us anything![40] A few days ago Anni held your photograph in her hands and then I noticed how she hugged and kissed it. We all long to see you again. Now [only] God knows whether it will one day occur. It is not impossible! After all there are still normal relations with the USA. How is it going for you? What are you doing? Do you have employment? Or, by now must you starve? We already have 10 chickens, also one piglet. If it survives, then we will have something to butcher in the coming autumn. Will close! With our greetings, yours, who pray for you [PJ 1883]

[39] It seems that church services were still being held in spite of the various prohibitions against religion. In Gnadenfeld, the nearby administrative centre for the region, the church was officially closed in 1936, Jacob A. Neufeld, *Tiefenwege,* p.51.

[40] This is, I think, sarcasm about the communist propaganda that the soviet system supplies all the needs of its citizens so well. The word *must* is used more in the sense of *need not*. It may also be a ploy to get the letter past the censors, since the Soviet state was now demanding that its citizens refuse foreign aid because they had no need of it.

COMMENTARY

Aside from the reiteration of the difficulties of sending letters out of the USSR, this letter provides an interesting insight into the lack of success the Soviet prohibitions against religious institutions and practice had in Kontiniusfeld. Comments like *Sunday after Sunday she has to stay at home* seem to indicate that there were regular church services being held and that these were more than just one or two people meeting in private. To compare this religious activity with what went on elsewhere suggests that, perhaps, this was somewhat unusual in the peasant villages in the Ukraine in 1934 and may be a direct reflection of theology in the Mennonite community [particularly the notion of the priesthood of all believers].

In the book, *Execution by Hunger*,[41] the author describes the destruction of the church in his Ukrainian village by the party apparatus and the army in 1930. From that day on there were no more organized communal or group activities of a religious nature until the author's departure in June 1933. However in 1934 on a wider scale in various parts of the Ukraine during Easter celebrations, churches filled with worshippers also came under attack by the local Komsomolsk, who destroyed icons, pictures and other sacred material in the midst of the services.[42]

The church in Kontiniusfeld had been officially closed since mid 1931. In addition, to decrees forbidding religious worship extra pressure had been put on the Molotschna congregations by placing heavy taxes on them if they wished to keep their churches operational. Even if they had already paid the required tax, new arbitrary taxes were also suddenly demanded.[43] In Lindenau after the Pentecost service in 1934 the preacher was immediately summoned to Halbstadt, the nearest regional Soviet administrative centre.[44] After a wedding in mid-May the bridegroom was under threat of losing his work in the artel, because a preacher had been asked to perform the ceremony.[45] Nevertheless, one reads of baptisms in Nikolaipol and Schoenwiese. Since

[41] Miron Dolot (pseudn.), *Execution by Hunger*, p. 28-30.

[42] DB 23/03/1934, p. 3.

[43] *Ibid.*

[44] DB 27/06/1934, p. 4.

[45] DB 8/08/1934 p. 3.

neither occurred in the Molotschna colony these events reinforced the opinion that some of the other colonies had greater religious freedom.[46] In September a Thanksgiving service was held in Chortitza-Rosenthal and preacher Aron Toews was still conducting services and Bible studies in November.[47] However, in the Caucasus by October the churches were all closed and the preachers were suffering persecution.[48] Yet in Kontiniusfeld regular Sunday worship had continued one way or another until the present moment. Elsewhere house services and Sunday schools were still being held as well.[49] Clearly attempts to destroy religious faith were not completely co-ordinated or evenhanded through out the Ukraine.

I suspect that the Russian communists failed to differentiate between the hierarchical structure of the Orthodox Church and the more personalized practice of the Mennonites that made worship of the Divine a more individual responsibility. Since the use of icons and priestly authority was less central to the Mennonite faith, Soviet anti-religious policy based on an understanding of Orthodox practice would not be nearly as effective in destroying religious commitment among the Mennonites unless the attack was directly upon the individual believer. The closing of the "church" building, the destruction of the church bell tower, or the removing and degradation of sacred icons or the priest as described by Dolot might be more effective in eliminating religious faith if the central focus of the practice was a hierarchical ritual conducted once or twice a week by a priest. Although in somewhat similar fashion attacks on the preachers in the Mennonite villages certainly curtailed church services, private and family devotions consisting of bible reading, personal prayer and even the singing of a hymn or song expressing some aspect of one's Christian faith, were much harder to eradicate if they did not require physical icons or a particular "holy" man for such religious practice to continue.

[46] DB 20/06/1934, p. 5, 11/07/1934, p. 3, 23/03/1934, p. 3.

[47] DB 31/10/1934, p. 4; 19/12/1934, p. 5; MR 17/10/1934, p. 7.

[48] DB 24/10/1934, p. 4, 31/10/1934, p. 3.

[49] DB 23/03/1934, p. 3.

1934 06 03

Jetzt schon 4 Monat ohne Regen

Lieber Jakob, da wir Gestern nach langem Warten endlich mal wieder von Dir einen Brief erhielten, so will ich auch gleich antworten. Der Brief war vom, nun kann ich nicht mal schreiben von welchem Datum er war, nur dass es Nr. 4 war, das weiß ich. Aber Couvert war keines drin, auch Papier nicht, ich meine leeres und im Vorigen auch nicht. Wir sind, Gott Lob und Dank, schön gesund und wünschen Dir dasselbe von Herzen. Es ist hier auch so wie bei Euch, sehr trocken. Nun es hat aber schon in letzter Zeit paar Mal ein wenig geregnet, aber nur sehr wenig und uns fehlt schon ein großer Regen, da unser Gemüse noch nicht aufgegangen ist.

Heute, den 12.6.1934 will ich weiterschreiben. Wollte warten mit Schreiben bis es geregnet hatte, aber gestern erhielten wir wieder einen Brief von Dir. Es war Nr. 5 und nun muss ich dann doch schreiben, wenn's auch nicht geregnet hat, damit doch endlich mal der Brief abgeht. Also jetzt schon den 4. Monat ohne Regen, dann kannst Du Dir denken, wie traurig es bei uns aussehen mag. Onkel D. Schartner sagte in der vergangenen Woche schon zu mir, wie traurig, mal keine Kartoffeln, keine Bohnen, keine Zwiebeln, keine Rüben – wie traurig, nicht wahr? Im vergangenen Jahr, dann waren wir ganz froh, denn wir hatten den Keller voll Rüben, dann hatten wir, wenn auch kein Fleisch nicht war, aber wir hatten Kukurusbrot und Rüben, so auch Zwiebeln und Bohnen, aber in diesem

1934 06 03

Now the 4th Month Without Rain.

Dear Jacob, Since, after waiting for a long time, we finally received a letter from you, I want to answer immediately. The letter was dated; now I cannot even write from what date, I only remember that it was no. 4 that I'm sure of. But there was no envelope or paper; I mean empty sheets, in it nor in the previous one. We are, praise and thank God, very well and heartily wish the same for you. Here as there where you are it is very dry. Recently it has rained a little bit a couple of times, but only a very little, and we have need of a big rain, for our vegetables have not yet sprouted.

Today, 12.6.1934 I want to write more. Wanted to wait until the rains came. However yesterday we again received a letter from you. It was no. 5 and so I must write even if it has not rained in order that our letter will be posted. So now the 4th month without rain, and you can well imagine how sad things appear here with us. Uncle D.[avid] Schartner already said to me last week, "How sad, only small potatoes, no beans, no onions, no beets – how sad, isn't it?" In the past year we were very happy, for we had the cellar full of beets. Although there was no meat, at least we had cornmeal bread and beets, as well as onions and beans.[50] However,

[50] This letter seems to confirm the statement of Jacob A. Neufeld, *Tiefenwege,* p. 27 that the crop in autumn 1933 had been a very good one as opposed to what occurred in 1934.

Jahr ist es das Gegenteil. Denn unser Gemüse ist noch nicht aufgegangen und auch die Baschtan[iv] nicht. So dass wir besser gehabt hätten, wenn wir nicht gesät hätten. Es wird in dem kommenden Winter schlimmer bei uns sein als es war einundzwanzig auf die zweiundzwanzig und auch vorvorigem Jahr war. Und das macht nur die Dürre. So traurig haben wir noch nie in die Zukunft geschaut und wieder bitten um Hilfe bei Dir, ist einem einfach beinahe nicht mehr eigen, weil es bei Euch nicht viel besser ist, wie Du schreibst, dass bei Euch auch eine große Dürre ist. Nun hat Gott es nun mal beschlossen, uns durch den Hungertod von hier abzurufen, was wollen wir dann machen. Wir können uns den Regen nicht nehmen. Nur Beten können wir und das tun wir. Also ist uns nichts übrig, als abwarten was uns erwartet. Man denkt, es sollte doch auf die Bevölkerung einen Eindruck machen, aber nein, gar keinen. Es ist so, wie es in der Offenbarung heißt im K9.2,6,21.K.16.9,11 Blatt 2. Da sahen wir, dass auch Buße tun eine Gnade ist. Wohl dem, der Buße tun kann, denn ohne Buße

[iv] Arbusen

this year is the opposite. Our vegetables have not sprouted, and not even the watermelon. So it would have been better for us not to have seeded. It will be worse for us this winter, than it was in [the winter] of twenty-one to twenty-two, and also than the winter before last (*vorvorigen*) [probably 1932/33]. The only cause of this is the long drought. Never before have we peered into the future with such sadness. Once again to ask for help from you goes against the grain since it is not much better where you are. For you write that you also are suffering through a long drought. Well, should God have decided to remove us from here via death through starvation, then what can we do about it. We cannot draw down the rain. All we can do is pray, and we do that. So nothing is left to us but to endure what awaits us. One expects that it [such circumstances] should make an impression on the population, but no, not a bit of it. It [the situation] is similar to what is described in Revelations 9.2, 6, & 21[51] and 16.9, 11.[52] There we see that even repentance is a grace. Blessed is the one who is able to repent, for

[51] v.6: *In those days men will seek death and will not find it; they will long to die, and death will fly from them.* v.21: *nor did they repent of their murders or their sorceries or their immorality or their theft.*

[52] v.9: *men were scorched by the fierce heat, and they cursed the name of God who had power over these plagues, and they did not repent and give him glory.* v.11: *and cursed the god of heaven for their pain and sores, and did not repent of their deeds.*

keine Vergebung. Es macht auch bei uns nur einen kleinen Eindruck oder soll ich schreiben keinen. Vorigen Sonntag war bei D. Schartner in der Scheune Hochzeit. Am Tage wurden noch geistliche Lieder gesungen, aber des Nachts, dann wurde der ganzen Sache die Krone aufgesetzt. Gerh. Bartsh und Katja Wiens hatten Hochzeit. Und gestern Nacht starb Frau Heinrich Abr. Siebert im Krankenhaus. Und so geht es hier seinen Gang. Niemand fragt, wo es hinaus wird kommen. Als ob alles so sein muss. Heinrich geht es besser, aber empfiehlt sich sehr der Fürbitte. Es geht im so, wie in Lukas 7 u. 8 steht. Der arme Heinrich.... aber es war nicht sein Wille. Nun Gott weiß, wofür es gut ist. Aber bitte schreibe nicht an ihn. Es wäre sein Schaden, ich weiß aus Erfahrung, könnte Dir darüber wohl schreiben, ist aber besser nicht. Nun fragst Du, was Hans verdient. Am Schluss des Jahres kann ich Dir schreiben, bis jetzt haben wir keine Ahnung davon. Er gibt ja Schleuder nicht heraus, hin und wieder auch

without repentance [there is] no forgiveness. Even among us [a reference to the specifically Mennonite population] it [the drought] has only a slight impact, or should I say, none. The Sunday before last there was a wedding in D.[avid] Schartner's shed. During the day spiritual songs were still being sung, but in the evening the entire event reached the concluding highpoint. Gerhardt Bartsch and Katja Wiens were married. And last night Mrs. Heinrich Abraham Siebert died in the hospital. And so life proceeds along its path here. No one asks what will be the outcome of all this; as if everything must be this way. Heinrich [HP1911] is faring better, but very much requests intercession. He experiences what is written in Luke 7.8.[53] Poor Heinrich... but it was not his wish. Well God knows what good will come of this.[54] But please do not write to him. It would be to his harm; I know from experience; could certainly write you a great deal about that, but it is better not to. You were asking what Hans was earning. At the end of the year I can answer that. At the moment we have no clue about that. Skim milk generally has no importance, but once

[53] *For I am a man set under authority.*

[54] What actually transpired, according to my aunts, was that he contracted Typhoid fever and was eventually discharged from the army upon a doctor's diagnosis that he was no longer healthy enough to serve. He returned to Kontniusfeld and joined his brother in working in the collective cow barn. See also JusP 1922, *Mein Wegbegleiter Jesus Christus,* p. 12.

echte. Von Onk. Benj. es erhalten. Herzlich Danke schön! Werde jetzt schließen. Sollte mir noch etwas einfallen, schreibe ich noch. Verbleiben Deine für Dich betenden und uns Deiner Fürbitte empfehlenden, Peter Rahn. [PJ1883]

in a while there is whole... [milk].[55] Received [that] from Uncle Benjamin.[56] Hearty thanks for it. Will quit now. Should something else come to mind, I will write it. [add it on]. Remain yours, the ones who pray for you and request your intercession. Peter Rahn[57]

[55] This implies that he brought home a lot of skim milk since it seemed unimportant to the authorities, but whole milk was not as readily available.

[56] [B. B. Janz 1877]. JP 1910 says in his paraphrase of this letter that B. B. had sent them money.

[57] PJ 1883 has actually signed this letter! Given his reprimand to my father in 1931 for signing his name, this indicates that he had lost his fear of the authorities or simply didn't care any more.

COMMENTARY

The tone of this letter and the one that follows is markedly in contrast to the optimistic tone of the previous one [1934 04 28]. The optimism of the former resulted from the fact that the harvest of 1933 had exceeded expectations through out the Molotschna,[58] the efforts to collectivize everything seemed to have eased slightly and some people were returning from exile.[59] Even if there was a lack of meat, they had managed to store enough vegetables to eat through out the winter and spring of 1933 and they had been allowed to acquire a piglet, with the hope of butchering it in the fall. In May the family in Kontiniusfeld and people in the Molotschna seemed to have enough to eat, though as in the case of Anni, most people had very poor clothing and/or bedding.[60]

Once again however, the weather compounded the misery brought on by the drive to collectivize and mechanize agricultural practice. In Kontiniusfeld as elsewhere in the Ukraine the mild winter had turned into a spring filled with drought and high winds and heavy frosts at night. The winter wheat was being ploughed under because it was not growing.[61] In addition some parts of the country experienced a large infestation of beetles or locusts (*käfer*).[62] The resulting grain harvest was generally poor,[63] and the vegetable gardens, the mainstay of most families, never did sprout properly.[64] In addition, at least in the eastern Molotschna, there was an early frost.[65] Thus whatever did grow was likely to be stolen, making the tending of gardens almost useless. Law and the erection of watchmen's huts around the collectivized fields

[58] Jacob A. Neufeld, *Tiefenwege,* pp. 27-28. They were still harvesting in some parts of the Ukraine in February due to the very mild winter, MR 28/03/1934, p. 3, and at least, partly because it was a good crop.

[59] DB 20/06/1934, p. 5, 11/07/1934, p. 4, 19/09/1934, p. 4.

[60] Anni Rahn Wiens 1927 in 1934 04 02 had only *ein Kleidchen und das ist schon ganz beflickt, so das sie Sonntag für Sonntag zu hause sitzen muss.* See also, e.g., MR 7/03/1934, p. 7.

[61] See also the following letter. DB 30/05/1934, p. 4, 27/06/1934, p. 4.

[62] DB 8/08/1934, p. 3.

[63] In the Chortitza, DB 11/07/1934, p. 3, 19/12/1934, p. 5; in other colonies, MR 28/11/1934, p. 6; in the Molotschna DB 8/08/1934, p. 3, MR 25/07/1934, p. 5, 17/10/1934, p. 7.

[64] DB 19/09/1934, p. 4, 12/12/1934, p. 5; MR 17/10/1934, p. 7, *Letters from Susan,* p. 122.

[65] MR 17/10/1934, p. 7 and see next letter.

prevented any gleaning.[66] By the end of September no bread was available. No flour was available to be bought. The children did not receive any food at school and were starving. Even the horses and the chickens couldn't find enough to eat.[67]

These conditions forced the children to scour the woods and forests for mushrooms and berries to be dried for the winter. People who had any privately grown fresh food such as grain or pumpkins tried to process it into some kind of flour themselves at home rather than take it to any mill where it would probably be confiscated.[68] During this fall the local authorities and the GPU forbade and prosecuted any private trade in foodstuffs, something that had developed particularly in vegetables earlier in 1933.[69] Is it any wonder that, as in the following letters, most people anticipated another terrible famine and large numbers were leaving the Molotschna and moving elsewhere, most notably the Caucasus.[70]

[66] MR 17/10/1934, p. 7, DB 12/12/1934, p. 5.

[67] MR 28/11/1934, p. 6; DB 24/10/1934, p. 4; MR 17/10/1934, p. 7; DB 7/11/1934, p. 4.

[68] MR 26/12/1934, p. 6; 28/11/1934, p. 6.

[69] DB 24/10/1934, p. 4. For 1933 see the comments of Jacob A. Neufeld, *Tiefenwege,* p. 27.

[70] See also MR 17/10/1934, p. 7, 26/12/1934, p. 6.

1934 07 08

Habens ja schon mehrere mal erfahren müssen Kont.feld.

Lieber Jakob, da wir vor einer Woche deinen Brief erhielten, so will ich versuchen, ihn zu beantworten. Wir sind – Gott sei Lob und Dank – schön gesund und wünschen Dir dasselbe von Herzen. Wir sind schon in der Ernte. Aber gut geregnet hat es bei uns noch nicht, denn unser Gemüse ist noch nicht aufgegangen. Der Kukurus ist bis ans Kinn. Blüht schon auf Stellen, sind auch schon Ähren angesetzt, aber nur auf dem Gartenstück. Aufs Feld siehts wegen der Dürre anders. Nun hoffentlich gibt's auch da etliche Ähren, wenn es einen durchentzenden Regen geben wird. Erbose ist auch noch nicht aufgegangen, dagegen die Kartoffeln ja. Aber ob wir die Aussaat erhalten werden, das hängt wieder von gutem Regen ab, wenn wir den erhalten. Bis heut war er nicht. Das Korn im Weizen ist sehr gut, das wird sich rechtfertigen. Gerste und Hafer ist eine Frage. Nun wir essen uns noch fett. Was kommen wird, werden wir ja erleben. Wir habens ja schon mehrere mal

1934 07 08

Have experienced it several times already Kont[inius]feld

Dear Jacob, Since we received your letter more than a week ago, I shall try to answer it. We are – praise and thank God – healthy and heartily wish the same for you. We are already into the harvest. But it has not had a good rain yet, for our vegetables have not yet sprouted. The corn [maize] is up to the chin. Is already blooming, some ears have already taken hold, but only in the garden plot. In the field because of the drought it is different. Now hopefully some ears will also grow there provided we have a good soaking rain (*durchnetzenden Regen*). The watermelon has not yet sprouted, but the potatoes, yes, but whether we will get enough for seedlings depends again on whether we will receive a good rain. Until today there has been none. The kernel in the wheat is very good. It will provide good returns.[71] The barley and the oats are a question mark. Well we are eating ourselves fat. What will happen, we shall certainly experience. We have had to experience it several times already. Like a year

[71] Thus in 1934 the summer wheat harvest appears to be better in Kontiniusfeld than in other parts of the Ukraine. DB 19/12/1934, p. 5. That this appears of little comfort to the family is because they have already seen several times, that the grain harvest never results in anything being left for the workers in the artel. It is all confiscated by the state.

erfahren müssen. So wie ein Jahr zurück, als wir Dir sehr dankbar waren und heute noch sind. Kopp Isaak war Abend zu Besuch. Seine Familie ist gesund, so auch die Familien der Schartners. Was die Schwester Dürks. Frau Derksens Schwestern betrifft, so kann ich so viel schreiben, dass Anna leidend ist, war vor etlichen Wochen hier in unseren Dörfern zu Besuch. Maria früher. Schicke mir doch die Adresse von Johann K. Derksen, seine Mutter hat die nicht. Wunderbares Wetter haben wir hier, am Tage sehr warm und des Nachts sehr kühl. Vor einer guten Woche hatten wir eine Nacht Frost, so dass der Kukurus im Garten verfror, so auch die Kürbisse. Werde für heute schließen, das nächste Mal mehr. Verbleiben Deine für Dich betenden und einer Fürbitte empfehlenden Eltern und Geschwister. Heinrich und Lena gesund. Dieselben. [PJ 1883]

ago when we were very thankful to you[72] and still are. Isaak Koop (Kopp) was here for a visit tonight. His family is well as are the Schartner family. Concerning Sister Duerks., Mrs. Derksen's sisters, I can only say this, Anna is suffering. She was here in our village for a visit a few weeks ago; Maria earlier.[73] In any case, send me the address of Johann K. Derksen.[74] His mother does not have it. We have curious weather here, during the day very hot and at night very cool. Just over a week ago we had a frost so that the corn [maize] in the garden froze as well as the pumpkins.[75] Will quit for today, next time [will write] more! Remain your parents and siblings who pray for you and plead for yours [prayers]. Heinrich and Lena [are] healthy. As always. [PJ 1883][76]

[72] PJ 1910 must have sent them Canadian dollars according to my aunt, JusP 1922 *Mein Wegbegleiter Jesus Christus,* p. 12. Clearly the winter of 1932/33 was the worst thus far in our relatives' experience of collectivization.

[73] According to my aunts, these appear to be the relatives of Aron Duerksen who lived next door to the East of the PJ Rahns. From earlier letters he had a reputation for laziness. See 1930/09/15 & 1930/10/0?.

[74] This was a relative of the Derksens who lived next door on the West side.

[75] The late frost was quite wide-ranging throughout at least the Eastern Molotschna. MR 17/10/1934, p. 7.

[76] Written by PJ 1883 according to the aunts but contradicted by JP 1910's notes that ascribe it to Aganetha Schartner. I think it is PJ 1883, because the signing off is characteristic of him. Moreover, Aunt Lena Rahn Pauls 1914 seemed very sure she could recognize his *beautiful handwriting*.

COMMENTARY

In the middle of this letter PJ 1883 guardedly writes about what he anticipates will be another winter of famine. Instead of mentioning famine and starvation directly, he uses sarcasm [*we will certainly be eating ourselves fat*] and generalizations [*what will happen* and *experienced before*] to convey his concern and then confirms the intent by referring to a past condition [*Like a year ago when we were very thankful to you and still are*] which would certainly be known to the addressee.

This writing *durch die Blumen* [through the flowers] was designed to get the letter past the censors while conveying the information indirectly. In most of the letters to JP 1910 from the 1930s there is very little of this kind of conversation. Here it is a reaction to the strong explicit prohibitions of the Soviet state in 1934 against using the words starvation and famine in letters to foreign destinations.[77] So concerned were the authorities about word of any impending famine getting out to the West [particularly America] that even envelopes sending mail to America were no longer made available at the post offices.[78] Another tactic to avoid the local censor in particular was to deliver the letter oneself directly to a train station.[79] That may be the implication behind statements such as, *so I must write even if it has not rained in order that our letter will be posted* [1934/06/03] or *Since there is an opportunity to pass this letter along... It is better to post the letter on one's journey.* [1935/03/29]. Clearly there was a timing issue involved in writing these letters that probably had to do with when and where it was posted. Finally, in the next letter [1934 10 07] PJ 1883 mentions that they will use some form of registration (*einschreiben lassen*). I believe this is the first mention of this tactic and shows how concerned they were to maintain contact with JP 1910.

[77] These restrictions on what could be written had resulted in a significant decrease in the flow of mail out of the Soviet Union. According to one writer, she wrote 18 letters to Canada in 1934 and only one reached its destination. DB 9/01/1935, p. 5.

[78] Neufeld 1999, p. 248 dates the increased restrictions to the ascendency of Hitler in Germany in 1933. This explains why there are no letters from late 1932 and most of 1933.

[79] The letters in MR 17/10/1934, p. 7 and *Letters from Susan*, p. 125 refer to some of the tactics used to get around the local censors. The reference to envelopes is perhaps for registered letters. Envelopes were already an issue in 1930. See letter 1930/12/22.

1934 10 07

Ziehen viele nach dem Kaukasus.

Lieber Jakob, da wir Gestern einen Brief von Dir erhielten, den vom Sechsten, so wollen wir gleich zurückschreiben. Haben uns herzlich gefreut über den Brief. Von all dem, was Ihr dort im Geistlichen genießt, ist hier alles tot. Manch ein Herz leidet darunter und geht zugrunde. O, möchte der liebe Gott uns bewahren und Gnade geben um nicht mitreißen zu lassen. Denn die Verführungen sind sehr groß. Es gibt viel Schweres durchzukämpfen. Sind oft dankbar, dass wir uns noch im eigenen Haus dürfen niederlegen. Und Heute noch Brot haben, ob wir mit ausreichen werden, wissen wir nicht. Kartoffeln haben wir nur wenig bekommen. Der Regen kam zu spät. Hatten ein nettes Schweinchen, das schlachteten wir Freitag, weil nichts zu füttern ist. Es schlachten schon viele wegen Futter halber. Hätten wir es füttern können, dann wäre es ein schönes Schwein geworden. Hatten auch 18 Enten, auch die müssen wir schlachten, auch für Hühner ist nicht Futter. Das bisschen Kukurus müssen wir zu Brot halten. Haben auch noch ein Ferkel, das möchten wir gern durch den Winter bringen. Von hier aus den Dörfern ziehen viele nach dem Kaukasus. Papa hatte auch schon Lust, aber

1934 10 07

Many are moving to the Caucasus

Dear Jacob, Since we received a letter from you yesterday, the one from the sixth, we want to write back immediately. Have truly rejoiced over that letter. Among all that it contained – the spiritual nurture that you are enjoying there is completely dead here. Many a heart is suffering as a result and ends up shattered. Oh, may our beloved God protect us and give us grace not to be dragged along. For the temptations are very great. There are many difficulties that must be struggled through. Are often thankful, that we are still able to lie down in our own house, and that we still have bread [food]. Whether it will last [until the winter is past] we do not know. We have received only a few potatoes. The rain came too late. Have a young pig, but had to slaughter it on Friday because we have nothing to feed it. Many are slaughtering their pigs because of the [lack of] feed. Had we been able to feed it, it would have become a very fine pig. Also have 18 ducks, but will have to butcher those as well, not even for them is there feed.[80] The small bit of corn [maize] we have to keep for food (Brot). Still have a piglet, which we would like to maintain through the winter. From the villages around here many are moving to the Caucasus.[81] P.[apa] was also interested if only

[80] Also mentioned frequently at this time in other people's letters, e.g., DB 7/11/1934, p. 4.

[81] Especially from the Molotschna, *ibid.*

wenn es nicht zum Winter ginge. Onk. Joh. R.[ahn] ist hingefahren besehen. Dann soll die Familie hinter an kommen. Aber weil da so viele hinziehen, bringt es da auch schon viel Schweres mit sich. Es sind nur meistens solche, die hier nicht ihr eigenes hatten. Es ist schwer zu handeln. Wären wir dann gezogen, als P. da war, dann hätten wir gut getan, aber der Mensch denkt und der liebe Gott lenkt. Onk. J. hat geschrieben, die Familie soll auch mit hin kommen. Papa ist Maurer. Hans Hirte. J.[usta] geht in Gnadenf.[eld] zur Schule, muss alle Tage gehen. Wie es im Winter wird, wissen wir nicht. Produkte

winter was not approaching. Uncle Joh[ann] R.[ahn][82] drove there to look it over. The family is supposed to follow. But because so many are moving there, many difficulties are also arising there. It is mostly those who do not have [anything of] their own here [that are moving there]. It is difficult to do any business. Had we moved when P.[apa] was there,[83] then we would have done well[84] but the human being imagines [dreams], but God directs. Uncle J.[ohann] has written that the family [ours] is also supposed to move there. Papa is bricklayer,[85] Hans [JohP 1913] a shepherd. J.[usta] is attending school in Gnadenfeld. Has to walk every day.[86] How that will turn out in winter we do

[82] Johann Jacob Rahn 1879, from Sparrau.

[83] This reinforces what we have noted earlier about the winter of 1932/33. See commentary to 1934 02 27.

[84] [Financially]. The point is that in 1932 private trade was still going on to a much larger extent than at the present moment [1934] for by now the authorities had clamped down much harder on all free enterprise. E.g., DB 24/10/1934, p. 4.

[85] It appears that there was an increase in building activity in 1934 through out the Molotschna as the confiscated larger houses had been taken apart to provide building materials for various Soviet projects. MR 25/07/1934, p. 5.

[86] Approx. 4 kms.

hingeben geht nicht, das kommt zu teuer. Zu Hause bleiben dürfen sie nicht. M.[ika] geht hier zur Schule. Ä.[nni], die ist noch bei mir zu Hause. Bespärrt mir auch schon Schritte. Aber wo ich hingehe, da kommt sie auch mit. Sie soll auch noch zur Schule gehen. Von Hei.[nrich] erhielten wir Freitag einen Brief. Wenn wir ihm können Reisegeld schicken, dann kommt er auf 10 Tage auf Urlaub. Lena ist Lungenleidend. Wir schreiben, sie sollte nach Hause kommen, aber sie schreibt, sie hatte gute Medizin und weil die Aussichten so schlecht sind, will sie noch dort bleiben. Nun möchte der liebe Gott ihr noch die völlige Gesundheit geben, denn für sie ist es dort besser. Sie sind da mit unseren zusammen. Lena Rei.[mer] von hier ist auch da. Die war hier auf Ferien, aber die fuhr wieder zurück. Das Leben gefiel ihr dort besser als hier. Wir sind – dem Herrn sei Dank – auch alle

not know. To provide produce is impossible,[87] it is too expensive. To stay at home is not allowed. M.[ika] is attending school here [in Kontiniusfeld]. A.[nni] is still at home with me. Already saves me steps. But wherever I go, she always accompanies me. She is also supposed to go to school [soon]. Received a letter from Heinrich on Friday. If we can send him travel money, then he will spend a 10-day leave with us. Lena is suffering from lung infections.[88] We wrote that she should come home, but she replied that she had good medicine and, because prospects here seemed so poor, she still preferred to stay there [in Leningrad]. So may God restore her health completely because for her it is better over there. They are together with ours [other Mennonites] there. Lena Rei.[mer] from here is also there. She was here for a holiday, but returned again [to Leningrad]. She preferred the life there to [ours] here. We are – thank the Lord – also all well. Only

[87] [To pay for board in Gnadenfeld with goods in kind]. This is rather curious, for since spring 1932 the children were supposed to eat at school at the state's expense. It may be that once again the state has reneged on earlier promises once it has achieved its aim and returned the responsibility for the children onto their families, now that it was apparent how difficult it actually was to accomplish all their collectivization goals at once.

[88] Aunt Lena says that this was actually an inflammation of the abdomen that resulted in large amounts of fluid being drained from her body over a period of 9 months, as much as 10 litres on one occasion.

gesund. Nur Pa. hat so Durchfall. Mi. und ich so Schnupfen. Nun, werde zum Schluss eilen, alles schläft schon. Nur P. ging Kranke besuchen. Großmamas Justa ihr Bruder, Peter F.[riesen], wird nicht mehr lange hier weilen, geht seiner Frau nach. Hatte P. sagen lassen, er möchte so gerne selig sterben. Bleiben dann 5 Kinder verweist. Nun, wünschen auch Dir all die beste Gesundheit und viel Gnade zum Segen für unser da zu stehen. Wenn Peters ein Bild erhalten, wo meine Schw. droben ist mit Kinder, da ist auch Just. u. M. oben.[ASR1889]

Da Mama schon geschrieben hat, so will ich auch noch paar Zeilen beilegen. Erhielten heute, den 10. von Lena wieder einen Brief. Sie schreibt, ist schon gesunder, dürfen um sie nicht besorgt sein. Gott sei Dank. Man hat auch ohne dem noch so viel, was immer das Leben schwer macht. Du fragst, ob wir schon

P.[apa] has a bit of diarrhoea.[89] M.[ika] and I have the sniffles. Well, shall hasten to close. Everything is asleep! Except P.[apa] went to visit the sick.[90] Grandmother's Justa's brother Peter F.[91] will not linger here much longer. Is following his wife. Had it reported to P.[apa] that he would really like to die a converted (*selig*) man. Five children will remain orphaned there. Now wish all the best of health for you and grace to be a blessing as our representative there. If Peters receive a picture in which my sister[92] with children is present, Just.[a] and M.[ika] are also in that picture. [ASR 1889]

Since Mama has already written, I also want to add a few lines. Received again today, the 10th, a letter from Lena. She writes that she is becoming healthier, that we do not need to worry about her. Thank God! Without that, one still has much that always makes life difficult. You ask whether we have

[89] The famine like conditions that had existed in 1932/33 and seemed to be arising again as the winter of 1934/35 approached resulted in several reports of dysentery or malaria. DB 27/06/1934, p. 4, 19/09/1934, p. 4; MR 25/08/1934, p. 6.

[90] These visitations, as indicated earlier, often consisted of PJ 1883 singing for the people he was visiting.

[91] This was, according to JP 1910's paraphrases and notes, a Peter Friesen.

[92] According to my aunt, Anni Rahn Wiens 1927, this was Maria Schartner. The photo was taken at the wedding referred to in 1934/06/03.

hungern, bis heute nicht, Gott sei Lob und Dank. Und wenn's uns bleibt, dann werden wir auch bis Weihnachten nicht hungern. Aber dann nach Weihnachten, daran mag ich nicht denken, da ist die Zukunft wie eine Nacht. Keine Hoffnung, als die von oben. Gott kann ja helfen, denn sein Arm ist nicht zu kurz, dass er nicht helfen kann, aber auch nur Er. Mag auch gar nicht an die Zukunft denken, tue es auch nicht, denn sie ist zu finster.... Werden den Brief einschreiben lassen. Nun sei noch herzlich gegrüßt von Deinen für Dich Betenden mit Timot. 6.6-8. (10.10.1934) [PJ 1883]

already begun to starve. To the present, not yet, praise and thank God. And if everything remains with us, then we will not starve until Christmas. But then after Christmas, about that I do not like to think, for there the future is like the dark night. No hope except that from above. After all God can help, His arm is not too short to help us, but only He. Don't like to think about the future at all, don't do it either, for it is too dark ... Will register the letter. Now hearty greetings from yours who pray for you with Tim 6.6-8.[93] (10.10.34).[94] [PJ 1883]

[93] *There is great gain in godliness with contentment: for we brought nothing into the world and we cannot take anything out of the world; but if we have food and clothing, with these we shall be content.*

[94] This date is in the handwriting of the last writer. Obviously the letter was not mailed for three days until PJ 1883 could take it to be registered, a new means of trying to ensure delivery.

1934 12 30

Bitte, bitte schicke nichts mehr

Lieber Jakob! Da wir an der Schwelle des alten Jahres stehen, mit al dem, was es uns gebracht hat, aber auch an der Schwelle des neuen Jahres sind, wo wir nicht wissen, was es uns bringen wird, so will ich noch in diesem Jahre '34 etliche Zeilen an dich richten. Zuerst wünsche ich dir Gottes reichen Segen im neuen Jahr. Und weiter teile ich dir mit, das unsere Lena, deine Schwester, dort in der ferne sehr krank ist, vielleicht ist auch schon nicht mehr unter den Lebenden. Wir erhielten von Lydia Driediger gestern einen Brief. Sie und Tina sind auch da, und die schreiben das Lena operiert ist. Man hat ihr sechs Liter Wasser abgenommen. Furchtbar krank ist das arme Kind. Was für eine Krankheit es ist, wissen wir noch nicht, denn die Ärzte konnten es nicht feststellen. Erst meinte man von Bauchfell-Entzündung und Wassersucht, dann noch etwas dabei, aber in den letzten Nachrichten ist nur von oben gesagtem

1934 12 30

Please, please send us nothing more

Dear Jacob: Since we are at the exit of the old year, with all that it has brought us, but also on the threshold of a new year, concerning which we are uncertain what it will bring, I want to write a few lines while still in the year '34. First, I want to wish you God's rich blessing in the New Year. In addition I wanted to inform you that our Lena, your sister, there in her distant home is very sick. Perhaps, she is already no longer in the land of the living.[95] We received this news yesterday in a letter from Lydia (*Lyda*) Driediger. She and Tina[96] are both also there [in Leningrad] and write that an operation was performed on Lena and they drained off six litres of fluid. Terribly ill is that poor child. What sort of illness it is we do not know, for the doctors have not been able to determine [that] for certain. At first it was thought to be an inflammation of the stomach [or digestive tract] and dropsy (*wassersucht*), then something else as well. However in the last report, only the former is

[95] This is my aunt, Lena Rahn Pauls 1914, who as of this date [Autumn 2009] is still alive and was able to describe her illness in detail. She had fallen and suffered internal injuries that resulted first in internal bleeding and later in a large build-up of fluid in her abdomen. After being hospitalized and then receiving treatment for nine months she was given a year's leave of absence [into 1936] from her post in Leningrad.

[96] The sister of Lydia according to Aunt Lena.

gesagt. Hein ist gesund geworden, er lag an Unterleibstyphus, ist gegenwärtig zu Hause auf zwei Monate. Den sehen wir noch mal, aber ob wir Lena noch einmal sehen werden, ist eine Frage. Als Lena krank wurde, schrieb ich ihr, sie sollte nach Hause kommen, und so ist es jetzt. Nun: Gott macht keine Fehler, sein Weg ist heilig! Was unser Leben betrifft, so sind wir, Mama und ich, so leidlich gesund. Essen uns noch alle Tage schön satt und werden es auch in der Zukunft, denn unsere Regierung gibt uns Brot heraus, und dann wird auch freier Handel sein mit Brot, so das

mentioned. Hein [HP 1911] has recovered. He was in bed with Typhus of the lower body. At present he is at home for two months.[97] Him we shall see again, but whether we will ever see Lena again is questionable. When Lena fell ill, I wrote to her, that she should come home, but she wanted to wait until the summer and come then. And now this outcome. Well, God makes no mistakes. His way is holy (*heilig*). Concerning our life – we, mother and I, are reasonably well.[98] Still daily we eat well and will also do so in the future,[99] for our government doles out bread to us and free trade in bread will also

[97] He never went back to serve in the army because he was given a medical discharge. Although he again worked in the MTS after JohP 1913 was put in the army [see letter 1935 03 29], it seems that at least for a time [probably while he was recuperating] he joined his brother in working in the collective cow barn. JusP 1922 *Mein Wegbegleiter Jesus Christus,* p. 12.

[98] (*Leidlich gesund*) [Literally – "healthy but suffering" or "suffering while healthy"].

[99] This sentiment is probably quite sarcastic, for there are numerous reports that the poor grain harvest resulted in the anticipation of another severe famine, for already in September there was no bread available for the workers in the Kolkhoz, and even the children in the schools were starving. MR 17/10/1934, p. 7, 28/11/1934, p. 6, 26/12/1934, p. 6.

ein jeder kaufen wird können, so viel er will, so das du uns kein Geld schicken darfst, wir können es doch nicht für unsre brauchen. Den Order haben wir erhalten, habe herzlich Dank dafür, aber bitte, bitte schicke nichts mehr, es hilft uns nichts, damit wir nicht in eine üble Lage kommen. Das Wetter ist kalt, und in der vorigen Nacht hat es sehr geschneit, so dass ein guter viertel Arschin Schnee liegt, der Weizen hat eine gute Decke. Justa soll heute aus der Schule kommen, hat sie auf vierzehn Tage frei. Von unserem "Nach-Onkel-David-ziehen" muss ich dir mitteilen, man lässt nicht.

Will schließen mit schreiben, verblieben deine für dich Betende. [PJ 1883]

be allowed.[100] So everyone will be able to buy as much as one wishes. Therefore you are no longer permitted to send us money. For we cannot make any use of it. We received the order [money order]. Thank you very much, but please, please send nothing more – it is of no help to us – so that we will not end up in evil circumstances.

The weather is cold and the night before last it snowed so hard that now there is at least 12 cms. (*1/4 Arshin*) of snow. The grain has a good blanket [of snow]. Justa is returning home from school today. She has a 14-day holiday. Concerning our "Moving-to-Uncle-David"[101] I have to inform you, it is not allowed.

Will bring my writing to a close. Remain the ones praying for you. [PJ 1883]

[100] The capricious nature of Soviet policy-making is once again reflected in this statement. During the fall of 1934 the Soviets had explicitly forbidden any private trade in bread and the GPU or local police severely punished anyone caught doing so. DB 24/10/1934, p. 4.
[101] This is code for moving to the Caucasus, something discussed in more detail in the previous letter, 1934 10 07.

COMMENTARY

This is the last letter ever received from PJ 1883 as noted by my father, JP 1910, on the front cover of the envelope. It left him in deep consternation for many years, for he always wondered if he had done all he could or whether he should have continued to send money to the family in Kontiniusfeld. Again the writer of this letter has returned to his old habit of not signing his name. That seems to suggest that he was again being pressured by the GPU. According to JusP 1922, *Mein Wegbegleiter Jesus Christus,* p. 12, around this time the people of Kontiniusfeld were forced by the *Polizei* [GPU] to write to their correspondents in other countries that they had no further need of foreign currency. By not signing the letter, but complying with the demand of the authorities to inform relatives that they had no need of financial help, PJ seems more concerned about getting the letter past the censors, for in the sentence immediately after he mentions receiving financial help and thanks JP1910 very much for it. Thus he is almost contradicting what he wrote in the previous sentence. If his main concern were to satisfy official demands, surely he would have signed his name so that it could be noted that he had complied.

Although requests for financial help continue to appear in the *Mennonitische Rundschau* throughout 1934, 1935 and even later, most of these are by people who have already been inventoried, arrested, exiled or simply evicted from their homes [frequently preachers]. Thus most of them have nothing to lose. However in late 1934 PJ 1883 had not been inventoried, he had surrendered his 16 desiatines to the kolkhoz, he was allowed to work in the Artel to earn wages, and the family continued to live in their small house. Thus, from their point of view, the family had much to lose if they should irritate the authorities and, particularly, the GPU. They saw it as in their interests to preserve what they had by obeying whatever injunctions the state imposed on them as long as it did not attack their dearest possessions, their house, cow, pig and small vegetable garden. As ASR 1889 wrote in previous letters, *We will sacrifice much to remain in our own home.*[102] What they

[102] See letters 1930 06 02, 1930 10 0?.

never seemed to have compromised is their faith in the fact that God was directing their lives.

For approximately another year PJ 1883 continued to live and work in Kontiniusfeld. However, according to Justa Prei Geddart, his daughter, in Spring 1936 he was one of the first men in the village to be arrested, jailed in Dnepropetrovsk, put on trial for listening to a radio at a mill operator's work place, sentenced to five years of confinement, three years' loss of civil rights and confiscation of one third of his estate, and sent to a work camp [Dolienka] near Karaganda.[103] Of course, the mill operator was the key witness against him. According to HP 1911 the GPU resurrected the old charge of his being a German patriot and spy because of his internment in Germany during WWI [Statute 54.10]. Although he could never be properly designated a Kulak, the Communist system was now justifying its persecution of its own citizens by resorting to the more general charge of *enemy of the state*. Then listening to a radio, particularily a German broadcast, could indeed be designated as a dangerous act and a crime.

In actual fact, he seems to have been better off in confinement than he had been in Kontiniusfeld. He was assigned the job of hauling food supplies from the city to the concentration-work camp [to a maximum distance of 40 kms.] using a horse and wagon. In addition, it was his duty to maintain the sauna for the authorities. As a result he received gifts of various kinds from them, which he may have shared with his family via a monthly letter that he was permitted to send them. In fact, he seems to have fared better in the camp than the family did at home where although they continued to live in their own house, they lost much of their furniture and possessions as a result of his imposed sentence. His work assignment reinforces what was already noted earlier, namely, that PJ 1883 had a record as one who could be trusted.

[103] JusP 1922, *Mein Wegbegleiter Jesus Christus,* p. 13, 14.

1935 03 29

Einen schweren Winter hinter uns!

Lieber Jakob!

Da es nun Gelegenheit gibt, den Brief mitzugeben, so will ich auch wieder gleich schreiben. Schickten vor einer Woche einen Brief ab an dich.[v] Es ist besser, den unterwegs abzugeben. Nun, ehe ich weiter schreibe, so wünsche ich dir die beste Gesundheit und das beste Wohlergehen in der so weiten Ferne; denn was gesund ist, lernt man erst in Tagen schätzen, wenn man krank ist, denn es sind Heute 14 Tage, dass Papa im Bett liegt und sich selbst nicht helfen kann. Erst Blindarmentzündung, Geduermentzündung, jetzt noch so starke Grippe, wir zweifeln an Typhus. Aber der Feldscherieza sagt nein, die Aussichten sind nicht Typhus; dann solche große Hitze. Essen nichts als Schleim vom Hafer und Gerstengrütze und Reis und Tee. Wenn er erst essen konnte, dann würde sich auch die Kraft mehr finden. Der Blinddarm ist auch noch nicht ruhig. Oft will die Hoffnung schwinden, ob der Herr uns ihn noch lassen wird. Wenn die Not so groß ist,

[v] Dieser Brief ist von JP 1910 nicht erhalten.

1935 03 25

A difficult winter behind us!

Dear Jacob,

Since there is an opportunity to pass this letter along [not to use the local post office], I want to write immediately even though I sent a letter off to you about a week ago.[104] It is better to post the letter on one's journey. Now before I write anything further, I hope that you enjoy the best of health and success in that so distant place. For one learns the value and meaning of good health most of all when one is ill. Today it is the 14th day that your dad (papa) lies in bed and is unable to help himself. First, he had an appendicitis attack and a bowel infection and now, in addition, a powerful flu (*grippe*). We are wondering if it is Typhoid. However, the local medic[105] says "no", the symptoms are not those of typhoid. Has a high fever and can eat nothing except watered down oatmeal, barley porridge, rice and tea. If he could only eat something, then his strength might increase. His appendix has also not yet settled down. Often we begin to lose hope that the Lord will allow him to remain with us. When things really seem

[104] The letter referred to is not among my father's collection. It probably did not get through the censor's perusal. Hence the importance of being able to post this letter from a different location. And, of course, this one did get through.

[105] (*Feldscherezia*). There was no local doctor in Kontiniusfeld, only a kind of paramedic.

dann weinen die Kinder so und wollen nur beten; und wunderbar bekennt der Herr, dass er da ist. Die Schmerzen geben nach. Sonntag schien es so, Papa würde Krämpfe im Gesicht bekommen. Darum musste Justel nach Kräutern gehen zum Abkochen und Legen. Wie nach Hause kam, sagte sie, sie hatte den Weg über gebetet, dass der liebe Heiland Papa sollte die Krämpfe abnehmen, und der Herr nahm sie ganz weg, ich kam schon nicht zum Abkochen. Jetzt ist die Temperatur schon ein bisschen gefallen, es war bis 39,9 hoch. Eine Nacht zog ich viermal die Wäsche um. Papa hat schon sehr abgenommen. Es war für ihn zu schwer in der Mühle. Und arbeitete noch, wie er schon leidend war, zu schwer und erkältet. Wenn Papa noch soll gesund werden, das wird noch lange dauern, bis er wird hergestellt sein. Nun, möchte der liebe Gott uns noch gnädig sein und uns Papa lassen, denn was soll ich mit 3 Mädels? Es ist gegenwärtig schwer, Papa ist krank und muss bedient werden. Es ging ihm einen Tag und eine Nacht Blut ab,

bleak, then the children cry so hard and want only to pray. And in wondrous manner the Lord demonstrates that he is still present. The pains decrease. Sunday it seemed that Papa was developing facial cramps.[106] So Justa went to gather herbs that were to be cooked and applied [probably as a poultice]. When she got back she explained that she had prayed the entire time that she was gone, that the loving Saviour would take away Papa's cramps. And the Lord took them all away. I never got to the cooking of the herbs. Now the fever has abated a little. It was 39.9 C. One night I changed the bedding four times. He has also lost a lot of weight. The work in the mill[107] was too hard for him. Moreover, even though already ill, he continued to work too hard and caught cold. If Papa is supposed to get well, it will take a long time until he is completely restored. Well, may our loving God be gracious to us and allow us to keep our Papa, for otherwise, what shall I do with three young girls. At the moment it is very difficult. Papa is sick and must be looked after. For an entire day and night he lost so

[106] [Bell's Palsy?]

[107] A flourmill. JusP 1922, *Mein Wegbegleiter Jesus Christus,* p. 16 describes the work in the local mill as requiring a great deal of strength, since it involved moving large sacks of grain on and off the scales by hand. Weakened by poor diet, it is no wonder that the work took its toll on PJ 1883.

das uns Angst wurde. Aber das ist weg. Hans musste Montag bis zum 1 Mai lernen fahren, 20 Werst ab von hier. Weil er in seinen jungen Jahren nichts gelernt hat. Nun kommt er dadurch vielleicht los. Er ist ja in der Stufe, wo Hein frei von ist, der ist jetzt auf dem Traktor, kommt nur zu Nacht nach Hause. Er sagt es geht so schlecht, er muss immer an zuhause denken. Kommende Woche, dann ist er am Tage zu Hause, dann wird es leichter sein. Es ist ein Glück, Justa ist diese Woche zu Hause, die hilft schon viel. Kam eben von der Mühle. Hatte Milch hingebracht. Hackt

much blood that we became very afraid. But that problem has now disappeared.

Hans began Monday, until May 1st, to take driving lessons[108] at a place about 20 *werst* from here because he had learned nothing in his younger years. Perhaps this will provide him with a means of escape.[109] He is now at the point from which Henry is now free. Henry is now driving tractor,[110] only comes home for the night. He says that it [the working conditions] is so bad that he must always remember home [presumably to keep his spirits up]. Next week he will be home during the day. That will make it much easier [for us here at home, with the father sick in bed]. Luckily, Justa was home this week. She is already a big help. She has just returned from the mill, brought milk there. Chops leaves and branches [for firewood]

[108] JohP 1913. The German phrase (*lernen fahren*) could also mean that he had to drive 20 werst [about 21 Kms] to improve his literacy skills.
[109] It is not clear from what, but maybe from service in the army as per the next sentence. For HP 1911 had just been released from service in the army because of his illness.
[110] He had returned to work in the MTS where he had worked prior to his military service. See note 60.

Strauch. Holen Futter für die Kuh. Heute haben sie und Mika den Garten umgegraben, andere Mädels halfen auch. Jetzt ging Mika Schneeglöckchen pflücken, es wird sehr gesät. Wir hatten einen anderen Plan. Papa wollte fahren, Hein auf dem Traktor, Hans auf der Steppe, ich draußen, und dann macht der Herr einen Strich, und dann ist die ganze Rechnung verdorben. Aber wie gut, dass der Herr einen Strich durch die Rechnung macht, denn wie weit können wir voraussehen. Von Lena erhielten wir vom 21 März einen Brief, die ist schon aus dem Krankenhaus draußen, auch schon fast gesund. Sie schreibt, sie hat

and some feed for the cow.[111] Today she and Mika dug the entire garden over. Some other girls helped as well.[112] Now Mika has gone to pick snow drops [a spring flower]; these are sown very plentifully.

We had a different plan. Papa wanted to drive;[113] Henry would take the tractor onto the steppes; I would work outside. But then the Lord draws a line through our calculations and all our plans are for naught. But how good that the Lord does cross out our plans. For how far can we actually look into the future?

From Lena we received a letter dated 21 March. She is already out of the hospital and almost fully recovered. She writes that she

[111] It is interesting to note that at this time, long after many villages had only collectivized stables, they still have their cow. This can be attributed to their careful adherence to fulfilling their milk quota, as can be seen in the previous sentence, and their willingness to surrender much else [like money] in order to keep their cow as ASR 1889 stated in a much earlier letter. e.g., 1930/12/2.

[112] JusP 1922, *Mein Wegbegleiter Jesus Christus,* p. 9-11 describes how at the Gnadenfeld Centralschule as early as age 10 or 11, she and a group of her friends from the village banded together to support each other in order to avoid joining the communist youth organizations. Now during this week of school break they undoubtedly continued to work as their own youth brigade.

[113] This may mean that in spite of PJ 1883's statement in the previous letter about not being allowed to move to the Caucasus, they were planning to go away secretly. No other journey is mentioned in the recent previous correspondence.

15,2 Kilo abgenommen im Krankenhaus aber jetzt auch schon vier Kilo aufgenommen. Sie schreibt, wenn sie nicht den Pass erhält, kommt sie nach Hause. Leichter ist es für sie dort, denn hier arbeiten wird sie nicht können; denn hier heißt es: durch alles durch, was es für Arbeit gibt. Einen schweren Winter hinter uns: erst Hein an Typhus, dann Lena so schrecklich krank und jetzt Papa. Aber es ja alles nicht von Ungefähr, der Herr hat unser noch nicht vergessen. Dunkel sieht die Zukunft aus, aber er gedenkt unser. Ich sagte, wie gut dass wir Papa noch pflegen können, er noch unter uns ist. Nun gedenket unser, es tut sehr Not. Grüsse alle die nach uns fragen. Und Onkel Johann Toews grüsse sehr von uns. Einen herzlichen Gruß von uns allen-Eltern mit Kindern. Mit Papa ist es Heute schon leichter. [ASR1889]

lost 15.2 kilos in the hospital, but has already gained back 4 Kilos. She also writes that if she doesn't get her pass[114] she will return home. It is easier for her over there because she will be unable to work here [as a nanny and household domestic]. For here the motto is "Mix up everything that is considered work" [or "Mess up everything that might provide work."][115] A difficult winter behind us! First Henry had typhoid, then Lena was so very sick and now Papa. However, all does not occur by accident; the Lord has not yet forgotten us. The future appears very dark, but He remembers us. I repeat, "How good that we are still able to look after Papa, that he is still with us."

Well, remember us! There is great need. And greet Uncle Johann Toews[116] heartily for us. And a hearty greeting to you from us all – parents and children. With Papa it is already easier today.[117]

[114] Lena Rahn Pauls 1914 says that this was a residence pass required of all those living in the Soviet Union. It had to be renewed after having lived in Leningrad for 3 years.

[115] (*Durch alles durch, was es fur Arbeit gibt*) She probably means that all structure to society is disappearing.

[116] Johann Aron Toews 1876, who had married JP 1910's aunt, Margaretha Janz 1880.

[117] This seems to be the last letter that JP 1910 received from his Stepmother for at least twenty years.

COMMENTARY

What is perhaps most interesting in the previous letter is what it does not say. It seems to explain that the winter of 1934/35 was difficult primarily because of illness in the family. Of course, that illness seems, at least in the case of PJ 1883, to be related to the digestive tract and probably poor nutrition. It is interesting to note that they still have a supply of oatmeal, barley, and rice in March supplemented by milk from their own cow and they still live in their own house. Thus the fears expressed in the last letter of 1934 about severe starvation may not have materialized to the extent that was anticipated.

Aside from the poor harvest, this fear for their lot in the coming winter had its origins in the increased effort of the authorities in the autumn 1934 to stamp out the private trade in foodstuffs.[118] To compliment this policy, people who received foreign aid, either money or relief packages, were being forced to surrender them to the authorities as *not needed*.[119] Moreover, those who received such foreign aid, even while they were starving, were required to write to their benefactors, as PJ 1883 had done in letter 1934 12 30, to request that they not be sent any more aid since they had *no further need of it because the government itself would take care of them*.[120]

What was implied by this statement was the introduction on 1 January 1935 with great fanfare of a *free trade in bread*.[121] This, however, was no free enterprise. Rather it seems to have been an attempt to replace the ration card system with a less bureaucratic means of dispensing food, especially bread, to the population. People could now buy up to two kilos of bread at officially sanctioned depots for varying prices with money they had earned

[118] DB 24/10/1934, p. 4.

[119] DB 6/02/1935, p. 6; MR 27/02/1935, p. 3.

[120] A standard phrase that reappears frequently in many letters written by different people in the Soviet Union, *ibid*.

[121] MR 2/03/1935, p. 5.

in the collective or had managed to obtain some other way. The bread was delivered each morning by train to the local station where it was then sold to the people of the region. In practice, people had to get up early in the morning to get a place in the long line-ups that awaited the daily arrival of the bread train. Since black bread was the least expensive, people purchased mainly this. The other forms of bread then had to be sold. So on subsequent days the only available bread would be the more expensive white or gray bread. Moreover not every village had a train station and thus a ready bread supply. Many a hungry person had to walk some distance to obtain this state provided food. And of course bread prices soon doubled.[122] At the same time there was now less food to be distributed in the schools or in the collectives.[123] Kitchens in the kolkhoz were often closed for the winter and only began to function again when the fieldwork began.[124]

In addition to the *free trade in bread* the authorities also began to return some of the corn that had been confiscated the previous autumn to the kolkhoz in the winter of 1935.[125] In spite of this and the other Soviet initiatives already mentioned one still reads of hard working, exhausted people supplementing their diet with field mice, crows and other organic vermin. Others are still requesting aid from North America.[126]

[122] MR 2/03/1935, p. 5; DB 20/02/1935, p. 5, 17/04/1935, p. 5.

[123] This was already the case in fall 1934. See letter 1934 10 07.

[124] In the previous letter DB 17/04/1935, p. 5.

[125] DB 6/02/1935, p. 5, 27/03/1935, p. 5, 17/04/1935, p. 5.

[126] DB 27/03/1935, p. 6. Most of these requests, however, come from people who have already been exiled, resettled or disenfranchised and so have nothing to lose.

Bruchstueck von einem Brief

Zu Hause...auch sehr traurig

1935 08 08[vi]

Zu Hause ist es in diesen Jahren auch sehr traurig. Die werden dort wieder sehr hungern. Viele haben jetzt schon kein Brot und alles ist dort so sehr traurig. Versammlung ist geschlossen und so keine Andachten, keine Bibelstunden mehr. Dort wird nur sehr gefeiert. U[nd] wie ist die Hochzeit. Es ist zum Weinen. Kata D. Wiens, die hatte mit Gerhard Bartsch Hochzeit, das sind Geschwister.[vii] Dort wurde Abends gespielt und getobt und hochgelebt. So geht es dort, aber wie viel mehr, die Ungläubigen, die gehen ganz zu Grunde. L. R., Schwester.

[vi] Den Datum hat mein Vater, JP 1910, hinauf geschrieben.

[vii] Mit Geschwister meint die Schreiberinn *Kirchenglieder*.

Letter Fragment

At home it is very sad

1935 08 08[127]

At home it is very sad in these years. They will hunger much over there. Many already do not have any bread and everything there is very sorrowful. Church is locked, and so no services, no bible studies. Only lots of celebrations over there! And how is the wedding? It is to be bewailed. Katja D. Wiens married Gerhard Bartsch. They are brethren.[128] During the evening they played, and caroused around and enjoyed the highlife. That is how it is there, but more so the unbelievers are going completely to ground. L. R., sister.

[127] These loose pages were included with the later letter from Tante Lena Rahn Pauls 1914 in the same envelope but do not seem to fit anywhere. This is probably a separate letter from August 1935 for on the back of the envelope, containing the letter from 1935 11 27, JP 1910 has written in German, "the 8th of August [1935 added later] I received this letter." Hence this must be a fragment of that letter, not part of the letter from 11-27 that is all one sheet and clearly dated. Moreover, JP 1910 went through these letters later and often put two letters and sometimes more of close proximity in time in a single envelope, probably to save space.

[128] i.e., members of the church.

Die Frieda ist Komsomolka und die kleine Musch, die ist wie Mika, die ist Pionier und dann noch viele Andere. Ich bin nur froh, dass ich hier bin. Geflucht wird beinahe jedes zweite Wort. Mädchen, die sind jung, und solches muss Jeder hören, so das Es dort nicht blühend ist.

Frieda is Komsomolskaya and the little yapper,[129] the one who is like Mika is a Pioneer, as are many others. I am only relieved that I am here. Almost every second word is a swear word. The girls [Justa, Mika, and Anni] are young, yet such [language] must be listened to, so that there it [the situation in the village, Kontiniusfeld] is not conducive to growth (*blühend*).

[129] These are apparently Reimer sisters that lived near the school [*Dorfschule*].

COMMENTARY

The news that the church in Kontiniusfeld was now locked reinforces the reports of the final church closures throughout the Ukraine. As the churches were closed, the buildings were often designated as clubs, theatres, and cinemas or, at best, as hospitals or schools.[130] The renewed push by the Soviet authorities to eradicate religious practice had begun in the autumn of 1934 with the usual practice of making a sudden assessment for excessive new taxes upon either the preachers and their families[131] or upon the congregation in order to renovate the building and make it fit for public use.[132] Moreover in the correspondence from the Ukraine that still appears in Mennonite publications there are repeated complaints that Sunday is always designated a workday.[133]

In June 1935 several preachers from the Molotschna colonies were put on trial for *distributing gifts from Hitler among the poor.*[134] The basis of this charge is probably some sort of aid that had been received via Germany for there are sudden renewed pleas in other published letters that future relief packages or financial aid not be sent via Germany.[135] The sentences ranged from six to ten years.[136] On October 6th the Chortitza Mennonite church celebrated its final Thanksgiving festival and its hundredth anniversary, only to be closed immediately afterward for failure by the congregation to pay the newly assessed renovation tax of 17,000 rubles.[137] To some extent this was a significant closure, since it was, if not the oldest, then certainly one of the oldest Mennonite churches in the Soviet Union. The response to these closings and arrests was that people who had fled to the larger cities began to attend Russian protestant churches and those who remained in the village

[130] Rudnerweide, Lichtenau, Alexandertal, Waldheim, Grossweide, Nikolaifeld, Ohrlof, Friedensfeld churches were all closed. DB 2/01/1935, p. 5, 16/01/1935, p. 6.

[131] See, e.g., *Letters from Susan,* p. 125

[132] DB 18/12/1935, p. 5. However, after their closure as places of worship they often seemed adequate for other public uses.

[133] DB 29/05/1935, p. 5.

[134] MR 26/06/1935, p. 2, 18/09/1935 p. 4.

[135] DB 18/09/1935, p. 6; *Letters from Susan,* p. 126, 128.

[136] MR 18/09/1935, p. 4.

[137] DB 13/11/1935, p. 5, 27/11/1935, p. 5, 18/12/1935, p. 5.

began to journey to nearby Russian villages for the same purpose.[138]

IN the autumn the authorities instituted a new round of inventories, confiscations, exiles, and/or resettlements that continued on into the spring of 1936. Although there is talk of ridding the villages of *kulaks*, most of the land had by now either been confiscated or surrendered to the collectives. This new series of arrests was part of a larger Stalinist purge of intellectuals and political and military personnel throughout the Soviet Union. It was in this wave of arrests that PJ 1883 was also caught up early in 1936 and sentenced to five years in a concentration camp near Karaganda.[139]

To compliment this new attack on what was left of the leadership the authorities announced a new directive in the Molotschna villages and perhaps elsewhere. Those men who had been arrested and sent to work camps would be released to live at home with their families if the families would leave their Ukrainian villages and settle in the areas of their spouse's confinement. This was probably an effort to populate some of the less inhabited regions of the Soviet Union where the government intended to exploit the natural resources. Needless to say, the Rahn family chose not to move from Kontiniusfeld. Instead PJ 1883 served out his sentence and returned to his home village early in 1941.

[138] MR 18/09/1935, p. 4, *Letters from Susan,* p. 130.
[139] JusP 1922, *Mein Wegbeleiter Jesus Christus,* p. 13. Anni Rahn Wiens 1927, *Gepriesen sei die Gnade Gottes,* p. 6.

1935 11 27

Das Du den Heiland gefunden hast...

Lieber Bruder, die Liebe Jesu sei in und bei Dir, lieber Jakob. Endlich mal nach so langem Warten und Sehnen ein Lebenszeichen gesendet. Es war eine Gebetserhörung, von Dir einen Brief. Danke sehr dafür. Hatte schon geschrieben und jetzt hatte ich die Adresse verloren, jetzt konnte ich keinen Brief schreiben. Und dann schrieb Großmama sehr schöne Briefe, habe mich so sehr gefreut und dann, dass Du den Heiland gefunden hast. Das war für mich die größte Freude, was es gab. Mit Freudentränen habe ich die Worte gelesen und kann dem Heiland nicht genug dafür danken, dass er uns alle zu seinen Auserwählten hat gemacht. Solch großes vorrecht zu haben vor so viele anderen. Besonders hier ist es so traurig und doch weißt Du Jakob, muss ich Gott danken, dass wir nicht zerstört werden. Und darum ganz besonders freue ich mich auch über unseren Bruder Hans, der ist in Kiev. Aber er schreibt

1935 11 27[140]

That you have found the saviour

Dear Brother, The love of Jesus be with you and in you. Finally, after such long waiting and looking, a token of existence has been sent. It was an answer to prayer to receive a letter from you. Thank you very much for it. Had already written, but had lost the address. So now I could not write [send] any letter. And then Grandma wrote such beautiful letters that I rejoiced greatly. And then that you have found the Saviour! That was for me the greatest joy that I could receive. With tears running down my cheeks I read those words and am unable to thank the Saviour adequately that he has made us all his chosen ones. To have such a great boon before so many others.[141] Here it is especially sad and still, Jacob, you should know that I have to thank God that we have not been destroyed. Hence I am especially glad about our brother Hans [JohP 1913] who is in Kiev.[142] He

[140] In 1935 the suppression of foreign correspondence was again intensified. For about six months Aunt Lena Rahn Pauls 1914 could still send letters to Canada from Leningrad, after none were allowed out of the Ukraine. Since letters could still be exchanged within the Soviet Union, for a brief period Aunt Lena became the only source of contact with JP 1910. That contact too was lost at the end of 1935.

[141] At this point the writer herself had just been baptized, in spite of the opposition to religion by the Soviet Government.

[142] He was now undergoing military training there as his brother, Heinrich HP 1911, had in 1933. He however

immer sehr schöne Briefe, so dass auch das Gnade ist. Er schreibt so oft "Lena bete für mich, ich bete für Dich" und sagt, dass er sehr viele Anfechtungen hat und nur Gott kann helfen. Deshalb schreibe ich auch zu Dir, hilf nur viel Beten, dann werden auch Heinz und Hans glücklich werden und auch alle anderen, ach wie schön wird's dort doch sein in Jesu Armen wirklich auszuruhen. Da wo die Sonne der Ewigkeit scheint. Ja lieber Bruder, wirst Du oft wohl mögen mutlos werden und klagen, aber wir müssen immer ausharren bis ans Ende. Ich habe auch hier nur eine dunkle Zukunft, aber nur mutig sein heißt es dann, denn ich sehne mich so auch nach den lieben Brüdern. Ich kann es mit Worten nicht beschreiben, wie gerne einmal in die Augen sehen würde. Aber wann wird das was werden, weist Gott. Weiß Du, schick doch, wenn Du mich noch als Schwester liebst, als Zeichen der Liebe ein Bild von Dir, wie es Dir sieht. Und dann freut es mich sehr, dass Du zur Bibelschule gehst, möchte Gott Dir Gnade geben, Ihm immer treu zu sein. Bete für uns und wir für Dich. Aus inniger Liebe von deiner für Dich betenden u[nd] Dich liebende Schwester, Lenchen Hoffe auf einen langen und sehr schönen Brief. Oft wird es Hier so mutlos Alles, aber wir müssen anhalten bis ans Ende.

always writes such beautiful letters, which is also a grace. He often writes, "Lena, pray for me, I am praying for you." He says he has many temptations and only God can help. That is why I am writing to you. Help me pray a lot! Then Heinz and Hans and also all others will become happy. Oh, how pleasant it will be over there to really rest in Jesus arms, there, where the sun of eternity always shines. Yes, dear brother, no doubt you will become depressed and lament, but we must endure (*ausharren*) until the end. I too have only a dark future here, but then one must be courageous, even though I yearn for my beloved brothers. I cannot put it into words how enjoyable it would be to look again into your eyes. But when that will be only God knows. You know, if you still love me, your sister, send me, as token of your love, a picture of your self as you appear now. And then [also] it pleases me greatly that you are going to Bible School. May God give you grace always to remain true to him. Pray for us and we for you. Love, from your sister who prays for you, Lenchen. Am hoping for a long and very nice letter! Often it becomes so hopeless here, but we must hold on [to our faith] to the end.

was not as fortunate. He was sent to Vladivostok where he served for two years, then was put in an internment camp in Khabarovsk and starved to death there during WW II. Apparently he married a Russian woman after his release from the army, but nothing more is known of her. JusP 1922, *Mein Wegbegleiter Jesus Christus,* p. 14.

COMMENTARY

This was the last letter that JP 1910 received from his relatives until after World War II. His sister Justa [JusP 1922] wrote the next preserved letter from a German refugee camp near Marburg on 13 April 1947. It makes reference to my father's ill health at the time, for he had just had a gallbladder operation. It is numbered No. 2. This indicates that there had been at least one previous letter, probably in the fall of 1946. I well remember the excitement with which my parents greeted me on my return home from school one day in the autumn of that year. Finally, a letter from one of my father's relatives had arrived from Germany. It was as if a great burden had been lifted from my father's shoulders. Although regular correspondence with the other relatives did not begin until the early 1950s, the first signs that some had survived the horrors of the previous decade meant a great deal to my father.

Much had changed since the last letter reproduced and translated above. Three of JP 1910's sisters have written their own memoirs[143] that provide a personal look at their experiences in the following years. Suffice it to say that all four of the sisters survived World War II, three of the four [Lena 1914, Mika 1924, Anni 1927] spent most of their working life in the Siberian and Northern work camps and cities of the Soviet Union, and all, together with their brother Henry [HP 1911] were finally reunited in West Germany in the 1970s and 80s. Two, Lena and Anni, are still living as of this writing. All five married and have families that all live in Germany.

This final letter no longer conveys much about the actual living conditions of the family in Kontiniusfeld. That is probably why it survived the censers' critique. It concentrates on the personal and spiritual life of the writer and her brothers. It presents a sharp contrast to the official stance of the Soviet government described in the previous letter and commentary, for it shows that personal religion and faith were not as easily wiped out as the large institutional structures, icons and figureheads. In the end, of course,

[143] Justa (Rahn) Prei, *Mein Wegbegleiter Jesus Christus* (Backnang 1980); Anna Wiens, *Gepriesen sei die Gnade Gottes* (Siegburg, 1991); Helene (Rahn) Pauls, *Herr, Deine Wege sind wunderbar* (Heubach, 2004).

not only did the family survive the rigours of the Soviet system, but also the churches and their leadership re-emerged as important buttresses of faith in the various former states of the Soviet Union.

One of the most startling institutional affirmations that the Soviet system had been unable to destroy the Christian faith in spite of its resorting to the most destructive methods known to humankind, is visible if one views the city of Kiev from a tourist boat on the Dnepr. In the nineteen thirties and forties many of the large Cathedrals of the city had been entirely destroyed and, as per the usual policy, the bricks had been used to build structures required by the Soviet institutions. At the same time the Kiev skyline had been altered through the erection of large concrete statues and buildings of massive size. Until 1989 these dominated the horizon and declared the power of the Soviet state. However, since the fall of Communism, some of the large and beautiful Cathedrals have been rebuilt and stand in sharp and ironic contrast to the rather ugly masses of concrete that symbolize what Soviet ideology imposed on its member states. It is as if the Christian faith stands once again beyond the power of Soviet hostility.

Conclusion

Family name abbreviations used in discussion of the letters

ASR 1889	Aganetha Schartner Rahn 1889 – 1968
PJ1883	Peter Jacob Rahn 1883 – 1942
JP1910	Jacob Peter Rahn 1910 – 1991
HP 1911	Henry Peter Rahn 1911 – 1996
JohP 1913	Johann (Hans) Peter Rahn 1913 – 1944
HelP 1914	Helena Peter (Rahn) Pauls b. 1914
JusP 1922	Justa Peter (Rahn) Prei Geddart 1922 -2004
Anni P 1927	Anna Peter (Rahn) Wiens b. 1927

CONCLUSION

The fate of my grandparents and uncles and aunts in Kontiniusfeld from 1930 to 1935 and then in the rest of the Soviet Union later must leave even the most sympathetic observer of the Communist experiment in some perplexity. Though the letters leave no doubt that the family was opposed in principle to the Soviet government's collectivization effort, they also make clear that the Rahn family made every effort to fulfil all the expectations imposed on them during this period of excessive political and economic upheaval. Although they withheld their personal property from the Stalina Kolkhoz until Spring 1933 when they ceded it voluntarily, it should be noted that they never had any personal property to speak of until 1921, some four years after the Communist takeover of the Russian empire. Prior to that time they had simply worked and lived as *anwohner*, who made their living as hired labourers and lived in rented housing. In the autumn of 1930 in spite of the likelihood of starvation during the approaching winter they dutifully delivered their entire harvest, which they had worked so diligently to grow and reap, to the state.[1]

Even as they remained true to their faith and attended whatever church services were available to them, the younger children attended school, mother, father and teenage sons worked in the *Artel*, and they attended meetings of the village Soviet. Son Henry [HP 1911], who contributed to the industrialization goals by training and working as a tractor driver and maintenance person, later served in the army as required by the Soviet State.[2] John [JohP 1913] tried to enhance the food supply through cultivating bees, pigeons and rabbits[3] and later also served in the army. Even as the family worked industriously to provide for themselves within the bounds of possibility, they shared what they could with others in the village. In acknowledgement of their contributions they received an extra land allotment from the village Soviet to cultivate for their personal needs.[4] Moreover, whenever the village Soviet needed some one they could trust to

[1] See letters 1930 06 15; 1930 08 03; 1930 10 13, 14; 1930 11 13.

[2] Letters 1930 12 28; 1931 04 02.

[3] Letter 1930 03 31.

[4] Letters 1931 01 22; 1931 04 23.

work in the office or in the supplies depot, they called upon our grandfather [PJ 1883] to undertake tasks that required integrity.[5] Even the local Soviet council and the GPU never had occasion to identify them as Kulaks of the first or second class, for they never experienced exile or resettlement and resided in their own house until they left Kontiniusfeld on the great trek to the West with the German army in 1943. Time and again searches of their property turned up nothing of a contraband nature, for they had literally delivered all their production to the state.[6] They write straightforwardly in the letters that they must fall into line because *our loving God allows it to happen.*[7] Even when working as a watchman over the stores of the Kolkhoz, PJ 1883 refused to help himself to unauthorized foodstuffs. When I asked my aunts specifically about the possibility of feeding his family through taking some of the foodstuffs from the storerooms either for himself or others, they replied that he had discussed this possibility with a *brother* and had concluded that as a Christian he must fulfil the duty of his post and respect the dictates of the government.

The return on all this hard work and support of the new system was a long litany of dehumanizing realities. Day after day one saw the helpless children reduced to starvation.[8] Constant harassment and interrogation of one's fellow villagers and oneself and ultimately the fate of family members pitted one against another destroyed the psychological well being of the individual and the social fabric of everyday life.[9] To see one's fellow villagers' estates inventoried and then to see six months later arrest, resettlement, exile or shooting that led to the destruction or disappearance of many families and friends, engendered terror and was a constant emotional drain even on those who were spared that turmoil.[10] The extremely pervasive famine of 1932-33 that was partly the result of short-sighted policies based on an excessively

[5] Letter 1934 04 28.

[6] Letters 1930 10 ?: 1930 10 13; 1930 11 13; JusP 1922, pp. 4-5, 12-13.

[7] Letter 1930 09 15.

[8] Letter 1933 12 23.

[9] Letters 1930 05 18; 1930 06 02; 1930 06 15; 1930 11 13; 1930 12 02; 1930 12 15; 1930 04 01; 1931 04 23.

[10] Letters 1930 03 31;1930 04 04; 1930 04 27; 1930 10 14; 1930 09 15; 1930 10 13.

ravenous quota system which ignored the need for quotas for the workers' survival, seed grain, cattle feed and agricultural production, reduced the few producers' and workers' enthusiasm to despondency.[11] Even when something had been rightfully earned and received, one could never be sure that it would still be one's own the next day.[12]

The ultimate reward, moreover, for all the contributions that the working classes made to the new economic system was personal starvation. Somehow within this utopian ideology that was supposed to provide the workingman's paradise, the agricultural worker, at least, was forgotten. Instead of returning value for service, those who loyally obeyed the dictates of the state and without whose contribution the system would never have survived were reduced to begging for help and would have starved to death without the support of relatives and friends living in other countries and foreign aid groups.[13] But even that denigration of these key contributors to the system seems not to have been sufficient for the Soviet ideologues. Ultimately my relatives were forced to deny the existence of the very personal needs that the soviet state had failed to provide for them. Having neither adequate clothing, heating, nutrition, or medication for illness they were pressured to demand of my father, JP 1910, that he send them nothing more *damit wir nicht in eine üble Lage kommen*.[14] Surely, it is not surprising that a system that destroys those very people upon whom it is built should ultimately collapse. What is astonishing is that it lasted for some seventy years.

Would-be apologists for the Soviet regime have suggested that it was necessary to destroy the old system before a phoenix could arise from the ashes. Such is the justification often advanced for the excesses of the early 1930s that placed my ancestors in such jeopardy and killed some six million people living in the Ukraine. However, after five years of destruction,

[11] cf. letters 1930 06 15 & 1931 04 10.

[12] Letters 1930 06 02: 1934 02 27.

[13] Letters 1931 03 05; 1931 04 10.

[14] Letter 1934 12 30: *so that we will not end up in evil circumstances.* Of course within another year the writer was arrested on trumped up charges and sent away to a work camp near Karaganda [JusP 1922, p. 13] even though no more financial aid was sent to him from my father.

arrests, resettlements, exiles and deaths of the kulaks and the establishment of the collectivized economic order neither the Rahn family nor most other agricultural peasants in the Soviet Union experienced a better standard of living than that which they had had prior to the Communist revolution. Instead my grandfather was arrested and confined to a hard labour camp in 1936, then rearrested along with my uncle [HP 1911] and sent back to another internment camp where within months he died of dysentery and malnutrition in 1942. Another uncle [JohP 1913] was inducted into the army and died of starvation in a work camp near Vladivostok. Except for one aunt [Jusp 1922] who escaped to Germany at the end of World War II, the other family members spent the years after the war working in labour intensive jobs, saw their children deprived of educational or professional opportunities and were under constant psychological pressure from the KGB through interrogation and night time arrests. Their standard of living never again approached that which they had achieved prior to 1929. The point, of course, is that the Communist apparatus in the Soviet Union, even while paying lip service to the Marxist ideal of liberating the working people, failed them miserably. My ancestors certainly endured the ashes of destruction but they never experienced the promised phoenix of egalitarianism.

Select Bibliography and Abbreviations

DB	*Der Bote*. A weekly newspaper published by the Canadian Conference of Mennonites, Winnipeg, Manitoba.
Dolot	Dolot, Miron (pseudn.), *Execution by Hunger* (New York, 1985).
Dueck	Dueck, Ulrich, *Die Familie Rahn von Tiegenhof* (Grosshimstadt, 1986).
Dyck, Staples & Toews 2004	Harvey L. Dyck, John R. Staples, and John Toews, *Nestor Makhno and the Eichenfeld Massacre: A Civil War Tragedy in a Ukrainian Mennonite Village* (Kitchener, 2004).
Dziewanowski	Dziewanowski, M. K., *A History of Soviet Russia* (Englewood Cliffs, New Jersey, 1985).
Epp 1962	Epp, Frank H., *Mennonite Exodus* (Altona, Manitoba, 1962).
Fitzpatrick 1982	Fitzpatrick, Sheila, *The Russian Revolution* (Oxford, 1982).
Fitzpatrick 1993	Fitzpatrick, Sheila, "Ascribing Class: the Construction of Social Identity in Soviet Russia," *Journal of Modern History* 65 (December 1993) 745-770. Reprinted in Miller, Martin, Ed., *The Russian Revolution* (Oxford, 2001).
Fitzpatrick 1994	Fitzpatrick, Sheila, *Stalin's Peasants: resistance and survival in the Russian village after collectivization* (Oxford, 1994).
GPU	Internal security police in the 1930s, known today as the KGB.
Gorky 1975	Gorky, M. et al., eds., *The Great Proletarian Revolution* (October - November 1917) (Moscow, 1975).

Kroeker	Kroeker, Maria, *Ein reiches Leben* (Weinsburg, 1949).
McClellan	McClellan, Woodford, *Russia: A History of the Soviet Period* (Englewood Cliffs, NJ, 1986).
MR	*Die Mennonitische Rundschau*. A weekly newspaper published by the Canadian Mennonite Brethern Conference in Winnipeg, Manitoba.
Neufeld	Neufeld, Jacob A., *Tiefenwege* (Virgil, ON, 1949).
Neufeldt 2009	Neufeldt, Colin P., "Separating the Sheep from the Goats ..." *Mennonite Quarterly Review* 83 (April 2009) 221-291.
Neufeldt 1999	Neufeldt, Colin P., *The Fate of Mennonites in the Ukraine during Collectivization and the famine 1930-1933* (PhD Thesis, Edmonton, 1999).
Lena Rahn Pauls	Pauls, David & Helene, *Herr, Deine Wege sind wunderbar* (self-published, 2004).
Penner 1963	Penner, Horst, *Ansiedlung mennonitischer Niederlaender im Weichselmuendsgebiet von der Mitte des 16. Jahrhunderts bis zum Beginn der preussischen Zeit* (Weierhof, 1963).
JusP 1922	Prei, Justa, *Mein Wegbegleiter Jesus Christus* (manuscript, 1980).
Rempel 1973	Rempel, David G., "The Mennonite Commonwealth in Russia: a sketch of its founding and endurance, 1789-1919" *Mennonite Quarterly Review* 47 (Oct 1973) 259-308; 48 (Jan 1974) 5-54.

Schnyder–Spross — Schnyder-Spross, Dr. Werner, *Die Familie Rahn von Zürich* (Zürich, 1951).

Schroeder 1990 — William Schroeder, *Mennonite Historical Atlas* (Winnipeg, 1990).

Toews 1975 — Toews, John B., Ed., *The Mennonites in Russia from 1917 to 1930, Selected Documents* (Winnipeg, 1975).

Toews 1988 — Toews, John B., Ed. & Trans, *Letters from Susan* (North Newton, KS, 1988).

Anna Rahn Wiens — Wiens, Anna, *Gepriesen sei die Gnade Gottes* (selfpublished, 1991).

Maps, Photos and Documents

Schematic of Sparrau Layout 1930s
(From Anni Rahn Wiens 1927)

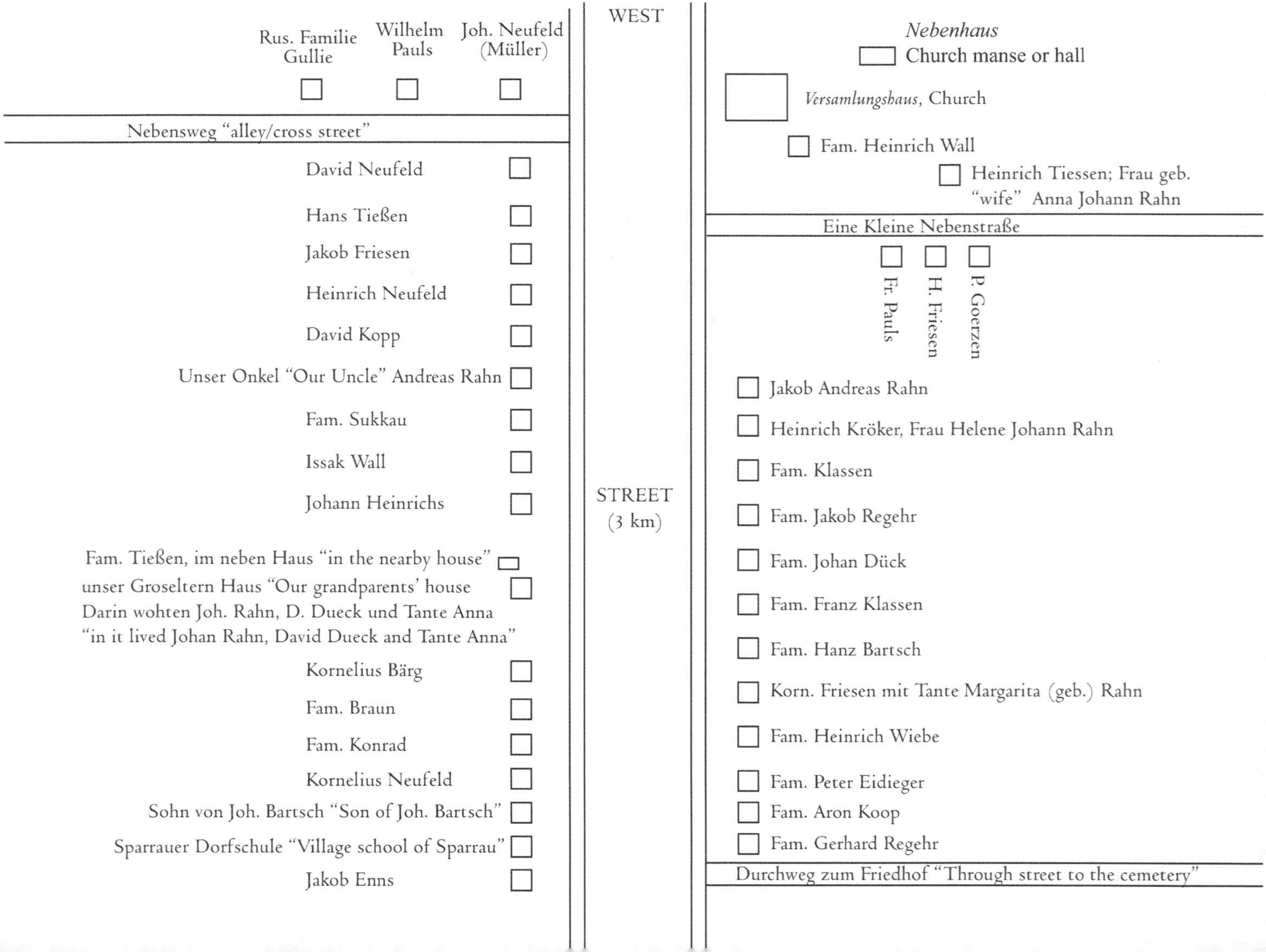

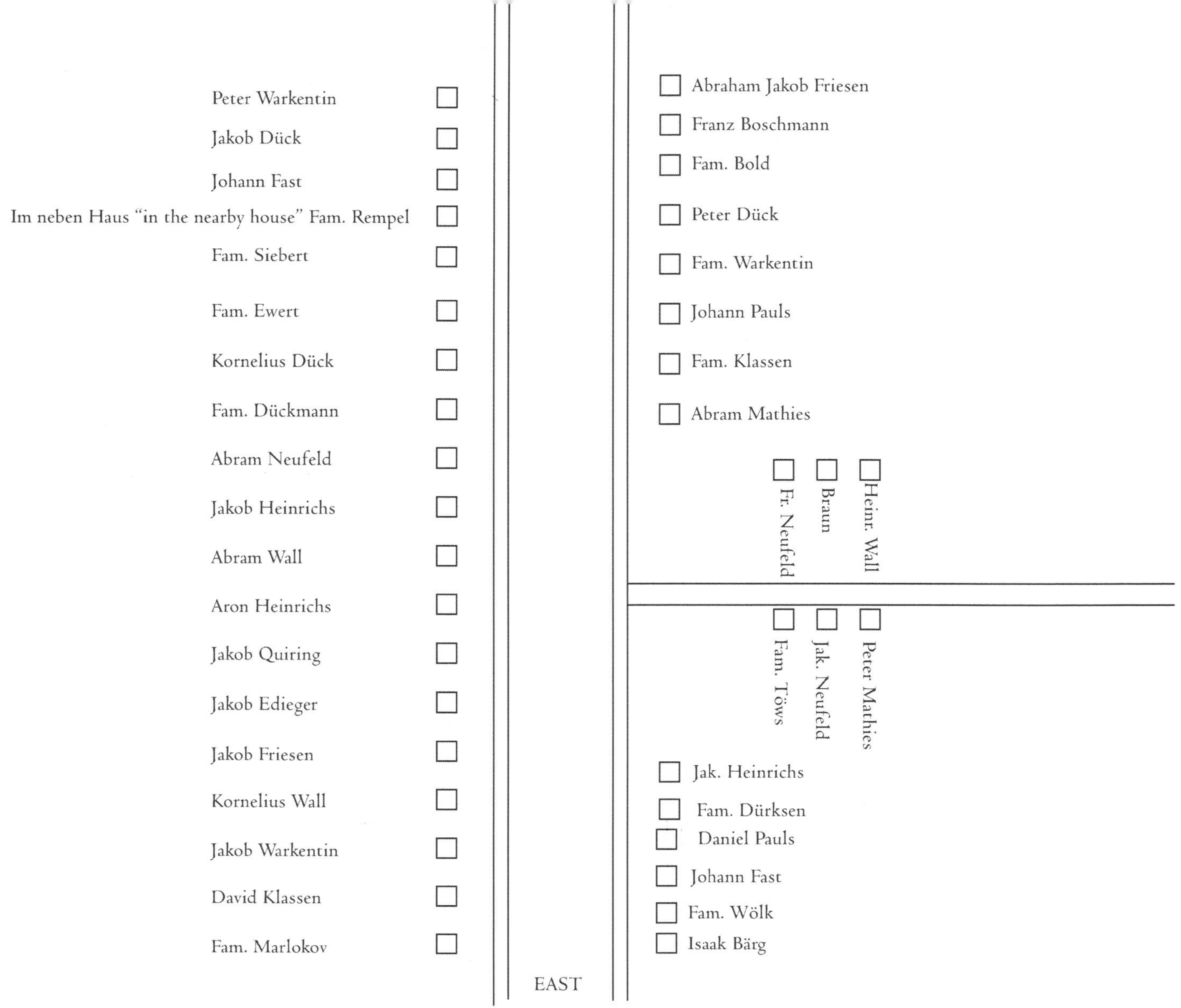

Peter Warkentin
Jakob Dück
Johann Fast
Im neben Haus "in the nearby house" Fam. Rempel
Fam. Siebert
Fam. Ewert
Kornelius Dück
Fam. Dückmann
Abram Neufeld
Jakob Heinrichs
Abram Wall
Aron Heinrichs
Jakob Quiring
Jakob Edieger
Jakob Friesen
Kornelius Wall
Jakob Warkentin
David Klassen
Fam. Marlokov
EAST
Abraham Jakob Friesen
Franz Boschmann
Fam. Bold
Peter Dück
Fam. Warkentin
Johann Pauls
Fam. Klassen
Abram Mathies
Fr. Neufeld
Braun
Heinr. Wall
Fam. Töws
Jak. Neufeld
Peter Mathies
Jak. Heinrichs
Fam. Dürksen
Daniel Pauls
Johann Fast
Fam. Wölk
Isaak Bärg

Diagram of Peter Jakob Rahn 1883 Family Property
(From Anni Rahn Wiens 1927)

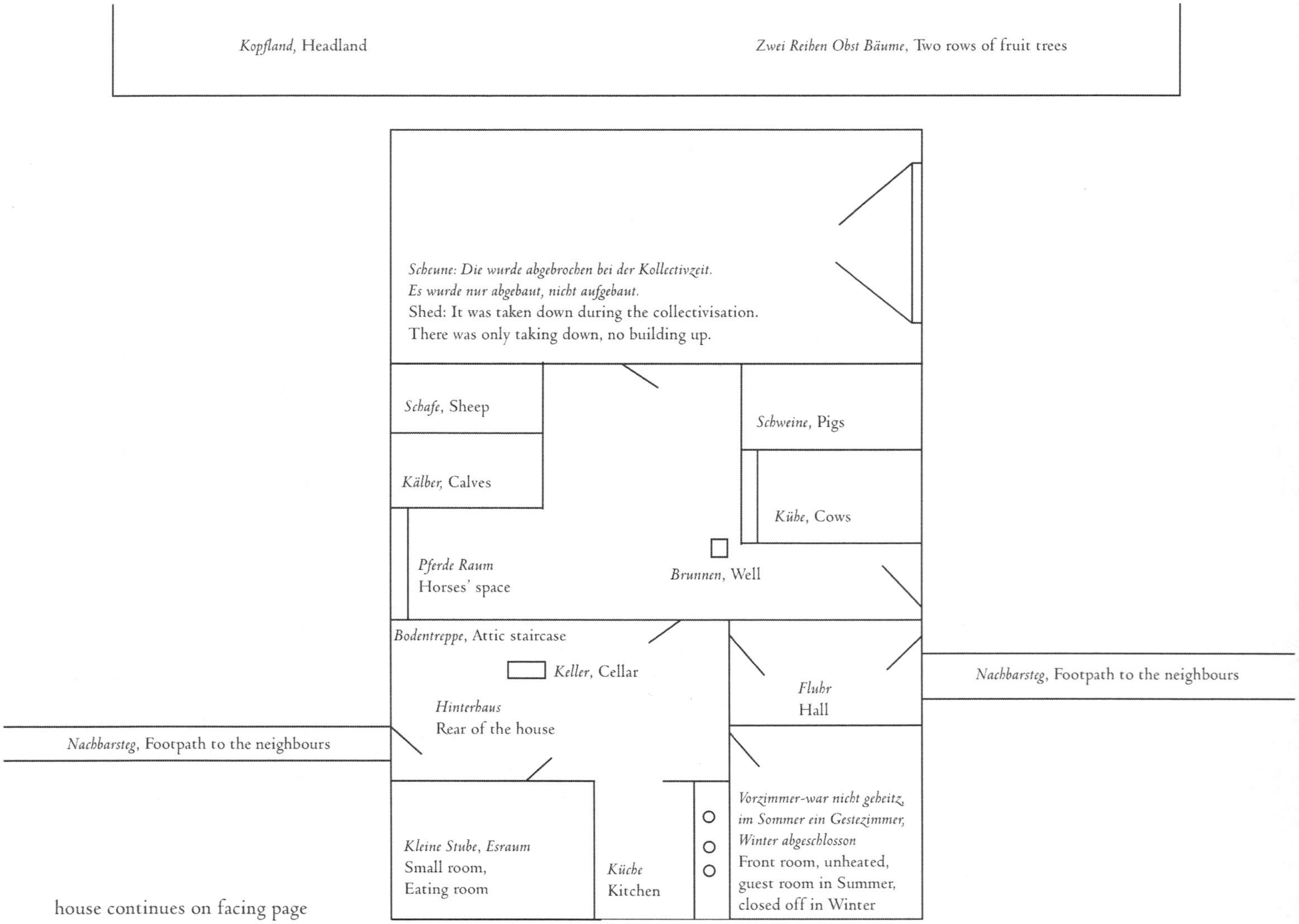

house continues on facing page

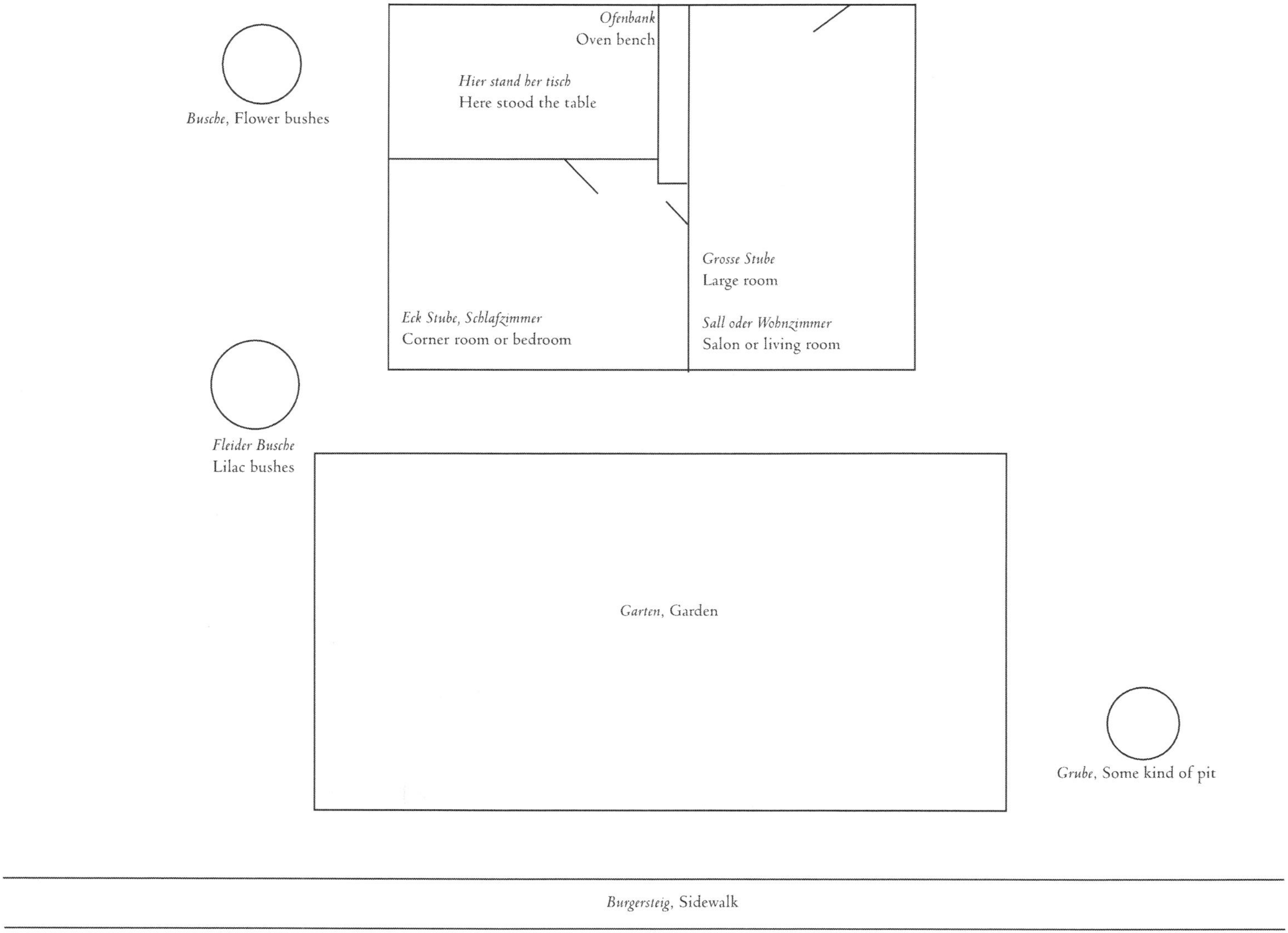
Busche, Flower bushes
Ofenbank
Oven bench
Hier stand her tisch
Here stood the table
Eck Stube, Schlafzimmer
Corner room or bedroom
Grosse Stube
Large room
Sall oder Wohnzimmer
Salon or living room
Fleider Busche
Lilac bushes
Garten, Garden
Grube, Some kind of pit
Burgersteig, Sidewalk

Schematic of Kontiniusfeld Layout 1930s
(From Anni Rahn Wiens 1927)

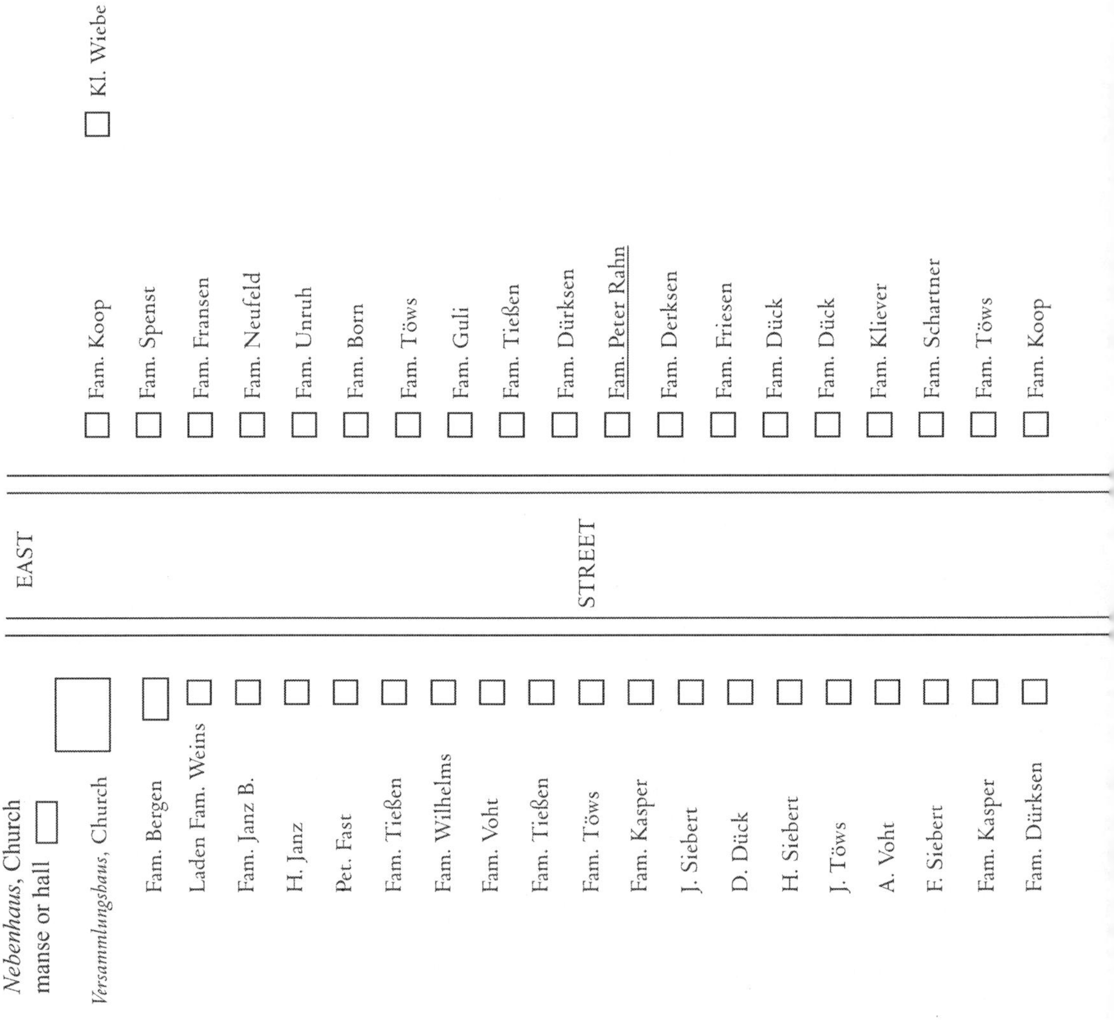

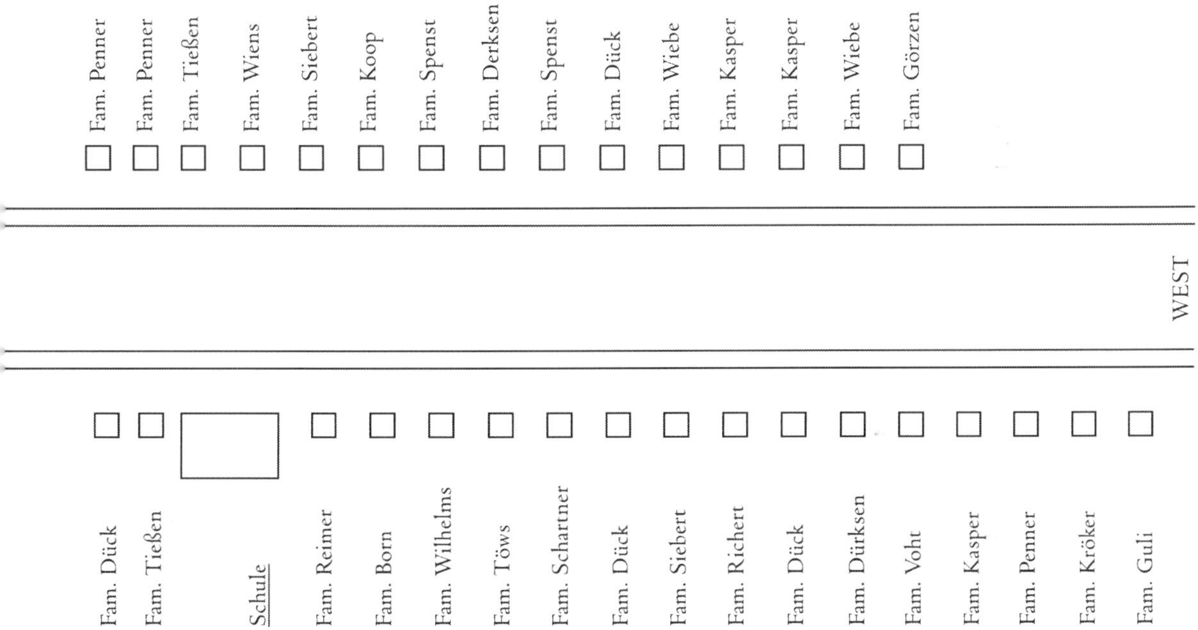
Fam. Penner
Fam. Penner
Fam. Tießen
Fam. Wiens
Fam. Siebert
Fam. Koop
Fam. Spenst
Fam. Derksen
Fam. Spenst
Fam. Dück
Fam. Wiebe
Fam. Kasper
Fam. Kasper
Fam. Wiebe
Fam. Görzen
WEST
Fam. Dück
Fam. Tießen
Schule
Fam. Reimer
Fam. Born
Fam. Wilhelms
Fam. Töws
Fam. Schartner
Fam. Dück
Fam. Siebert
Fam. Richert
Fam. Dück
Fam. Dürksen
Fam. Voht
Fam. Kasper
Fam. Penner
Fam. Kröker
Fam. Guli

Villages of the Eastern Molotschna 1930s

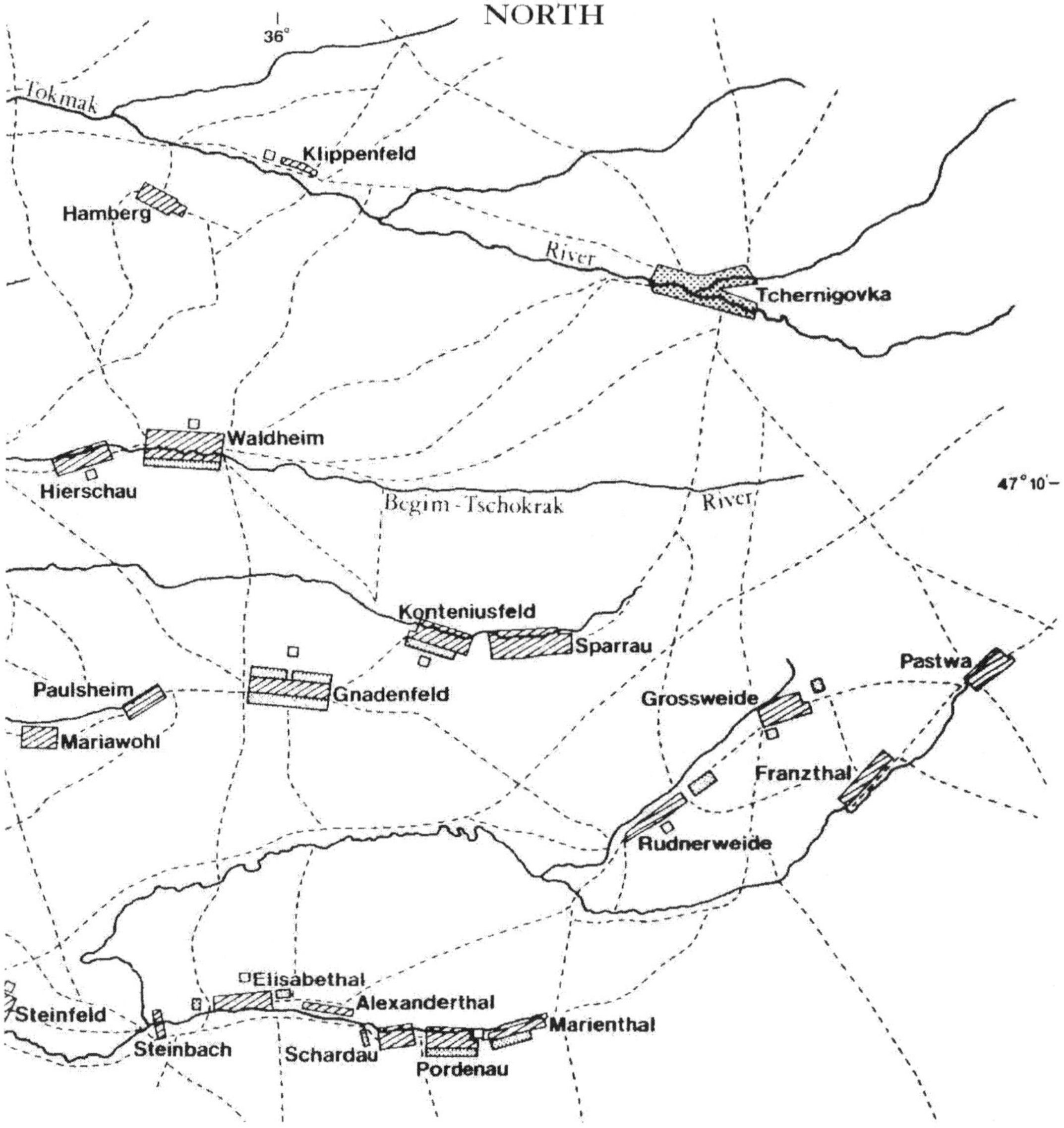

Back row (l to r): Abraham Walter Rahn, Henry Edward Rahn, Benjamin Arthur Rahn, John Daniel Rahn, Victor David Rahn, Peter Jacob Rahn
Front row (l to r): Mary Ann Suzanne Rahn, Mary Adeline Rahn, Jacob Peter Rahn, Helen Ruth Rahn

Peter Jacob Rahn b. 1883

Helena Peter Rahn, Maria Janz Rahn
Henry Peter Rahn, Johann Peter Rahn, Jacob Peter Rahn

Peter Jacob Rahn, Johann Peter Rahn?, Unknown
Agenetha Schartner Rahn
Mika Peter Rahn, Annie Peter Rahn, Justa Peter Rahn

Aganita ? Janz, Maria ? Janz, Helena Janz, Margareta Janz Toews, Maria Janz Rahn, David Derksen
Jacob Janz, Benjamin Janz Jr., Helena Vogt Penner Janz, Benjamin Janz Sr., Johann Toews, Peter Jacob Rahn

Tante Lena Rahn Pauls 1914 *Tante Anni Rahn Wiens 1927*

Die Beschreibung meiner Reise aus Russland im Jahre 1929 den 8/ten Nov.

Von zu Hause weg gefahren am Freitag 4 Uhr nachmitag im Jahre 1929 den 8/
ten/November, In Franztal angekommen um 6 Uhr abends, hier 4 Stunden gewesen
und um 10 Uhr abends weg gefahren zur Station. Auf der Station angekommen
um 11 Uhr Nachts, die Station heist: Nelgowka. Dann um 12 Uhr Nachts weg
gefahren nach Moskwa, dort angekommen am 10/ten November 1929 um 10 Uhr
des Morgens im Kurskij Wagsal. Von hier ging es zum Jaraslawskij Wagsal.
Hier kammen wir um 12 Uhr auf Mitag an und blieben hier biss 9 Uhr Abends.
Dann hatten wir ein QUatier gefunden fuer 35 Rubel den Monat. Das Quatier
war 100 Werst von Moskwa nach Norden in einem Russendorf: Listwjane. Im
Quatier angekommen um 11 Uhr Abends, es war ein hoelzernes Gebaeude.
Jetzt ging es zur Ruhe, wir grossen schliefen alle gut aber die Kinder
waren so unruhig das es hin und wieder mal einen Klapps gab aber das
wollte nur wenig helfen. Den folgenden Tage am 11/ten ~~Okto~~ November 1929 musten
wir uns inGruppen ein schreiben lassen und bei der Militz regestrieren.
Als diese beendigt war gingen wir uns Brotkarten hohlen. Auf der Brotkar-
-te bekammen wir 5/8 pf. das sind 300 gramm Brot auf den Esser. Das war zu
fuer einen Tag und kostete 11 kopeken. So gab es auch Zucker und Fleisch.
Als wir die Papiere in ordnung gebracht hatten und nach zu essen hatten,
wurden wir gewar das draussen die Roten schaften und drinnen nicht weni-
ger/. Jetzt war es doppelt traurig draussen fuhren die Autos und nahmen
in einer Nacht bis 200 Mann fest, schickten die dann zurueck oder im Keer-
ker hin ein. Aber drinnen war es noch schrecklicher, denn dieses Volk peinig
-te uns so das wir nich schlafen konnten. Und wer war dieses Volk, es war-
en die grausamen Peiniger: "Die Wanzen." Von Angst erfuellt vor den Inne-
ren und Auseren haben wir oft des Nachts unter dem Fenster gesessen und
gewartet ob die Reihe bald an uns sei. Es hat Geglueck, entlich kamm die
Nachricht das wir Auswandern durften. In dieser Zeit war benahe ein Monat
vergangen, noch eine kurze Zeit und in diesr Zeit erhielten wir den Pass
So das wir in einem Monat da weg fuhren. Am 5/ten Dez. 1929 bestiegen wir
wir den Zug um 5 Uhr Abends und als wir die Pässe erhalten hatte, war es
11 Uhr Nachts. Also um 11 Uhr Nachts fuhren wir los der Grenze Zu. Auf dem
Wege 200 Werst vor Sebesch (die Grenzstation) wurden wir nach gesucht, es
ging alles ganz gut ab. In Sebesch angekommen, standen hier ein und ein h
halb Stunden es War diese um 11 Uhr Nachts. Halb eins fuhren wir los
und dann ging es durch dem Roten Thor ueber die Grenze weg. In Lettland
kammen wir um 2 Uhr an. Auf der ersten Station musten wir umsitzem um 3Uhr
in lettische Wagone ein. Hier wurden wir sehr auf genommen. Zuerst kamm der
Arzt, der unter suchte ob da Kranke seien. Dann brachten die Schwestern Zu
Zucker, Kaffee, Milch, Bulka, Butter und Wareng, aber was machten wir?
Wir dankten fuer alle die guten Gaben, assen uns recht schoen satt und
gingen dann zur Ruh, schliefen dann wie auf weichem Moos auf einen Schanel.

Jacob Peter Rahn 1910 – Escape from Russia
Page 1

Der Zug ging schnell und am Morgen um 10 Uhr am 7/ten Dez waren wir 6
Werst vor Riga, dort hielt der Zug. Es gab hier Mitag und wir musten
noch vor dem Mitag um steigen im Deutschen Zug. Nach dem Mitag wude noch
eine Predigt gehalten und den lieben Gott gedankt fuer die Errettung.
Dann ging es los nach Littau und in Kowno gab es Abendbrot: Semmel mit
Wurst, Kaffee mit Zucker. und Milch. Dann gab es noch Schockolade mitder
Aufschrift: " Ein Gruss von den Deutschen Kownas ". Dieses war am 7/ten Dez.
8 Uhr Abends. Am 8/ten um 12 Uhr Nachts kammen wir uaf der Deutschen
Grenzstation "Eutqunen" an. Die Kranken wurden so gleich von den Arzt
ab geholt und dann musten wir uns zum Baden fertig machen. Unsere
Kleider wurden desenfexiert und wir musten uns baden. Nach dem baden kam
-men wir in ein zweistoeckiges Gebaeude. Hier gab es noch Kaffee. Am
8/ten Dez. Morgens gab es Kaffee und Weissbrot, zu Mitag Erbsensuppe, zu
Abendbrot gab es Tee, Semmel und Wurst. Von hier ging es um 7 Uhr Abends
weg nach Hammerstein. AUf dem Wege bekammen wir in Marihenburg noch
Tee an den sellben Abend. Dann gingen es durch Polen und am Morgen um 4
Uhr kammen wir in Hammerstein an. Es war am 9/ten Dez. 1929. Vom Zuge ging
es zur Aufnamebarake und da gab es Fruestick, Semmel, Wurst und Tee. Von
ging es nach einer ander Barake wo wir bis zum 14/ten Dez. blieben. Am 14/
Dez. wurden wir desenfegiert und kammen nach Haus N. 2. in Stube 26. Hier
lagen wir auf stroh bis zum 17/ten Januar 1930. Am 17/ten Januar 1930
Bekammen wir Bettgestelle mit Matratsen. Hier blieben wir bis zum 4/ten
Februar 1930. Am 4/ten Februar fuhren wir nach Moelln, uaf der Reise bekam
-men wir Ruebelplatz auf Neubrandenburg. Von der Station Moelln wurden
wir auf Autos ab geholt und in ein finfstoeckiges Gebaeude gebracht. Wir
kammen dann auf Fluhr-H. Stube 137. Um das Essen hier zu erlangen muste
wir in der Reihe stehe Hier blieben wir bis zum 27/ten Februar und fuhren
dann nach Hamburg. Auf der Fahrt nach Hamburg musten wir in Buchen umsitzen
von da gings bis zum Grossbahnhof Hamburgs. Von hier kammen wir zum Ueber
-seeheim. Dann musten wir am 28/ten vor dem Arzt der mich fuer gesund er
-klaerte. An densellben Tage muste ich auch noch vor den kanaedischen Kon
-solat der mir das Wiesieum zur Eireise gab. Am 6/ten Maerz teilte man
uns die Paesse aus und um 11Uhr auf mitag bestiegen wir den kleinen Dam-
pfer "Phoenik". Mit diesem fuhren wir zum grossen Dampfer "Melieta". Dort
stiegen wir sofort ein und dann gabs Mitag. Am densellbe Tage am 6/ten
Maerz um 5 Uhr abends fuhren wir auf der Ellbe los und um 10 Uhr Nachts
wurden die Anker ins Meer geworfen. Hier standen wir bis am 7/ten Maerz
2 Uhr nachmitag und dann gings uaf offener See. Und so fuhren wir auf
stiller See bis am 8/ten Maerz, um 10 Uhr abends kammen wir im Hafen Eng-
lands " Samtemton " an. Hier standen wir bis 8 Uhr morgens am 9/ten Maerz
fuhren dann nach Frankreich und um 5 Uhr abends waren wir dort. Um 7 Uhr
desselben Tages fuhren wir ab und dann gabs die "Seekranheit" - was mesra
-bles. Die war so gross das man zwischen Leben und Tod schwaebte, als ob
kein Erbarmen sei. Und so fuhren wir bis zum 17/ten Marz an welchem wir
in Hafen "St. John" N.B. An kammen. Von hier fuhren wir 5 Tage bis Coaldale.
In Coaldale angekommen am 22/ten Maerz 1930. Jacob Rahn.

Jacob Peter Rahn 1910 – Escape from Russia
Page 2

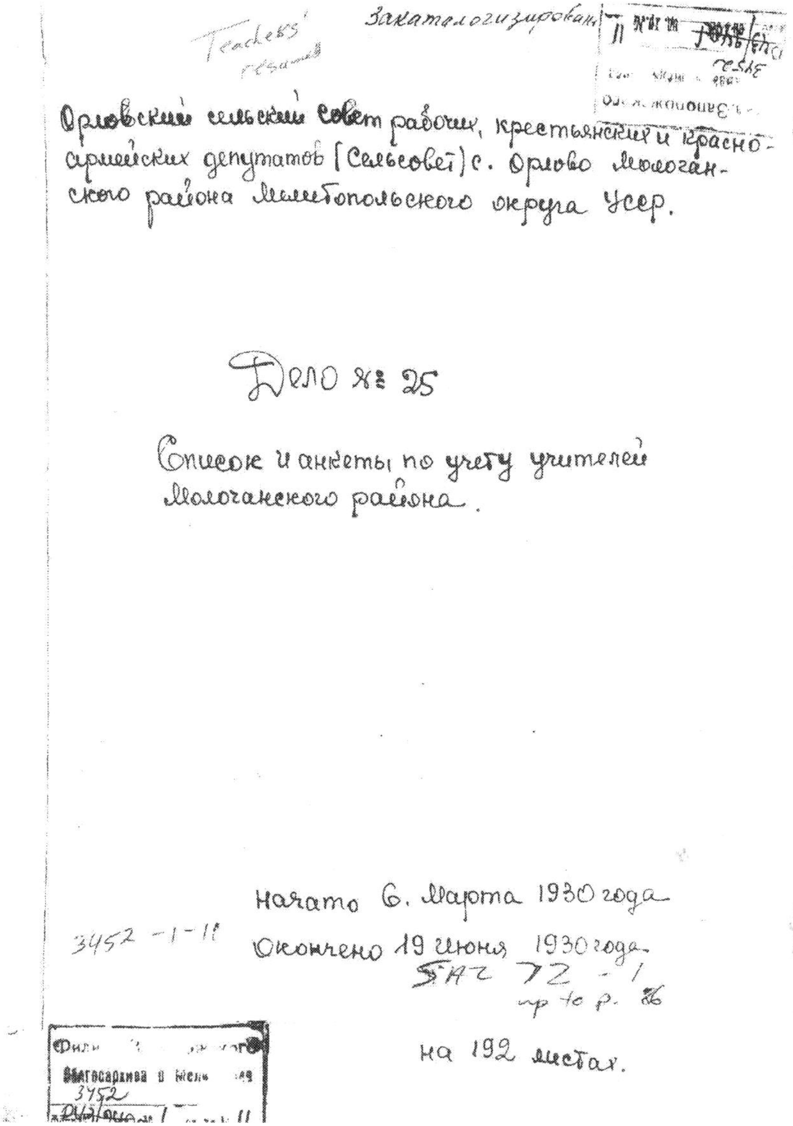

Teachers' resumes

Закатологизирован

Орловский сельский совет рабочих, крестьянских и красноармейских депутатов (Сельсовет) с. Орлово Молочанского района Мелитопольского округа УССР.

Дело № 25

Список и анкеты по учету учителей Молочанского района.

Начато 6. Марта 1930 года

3452-1-11

Окончено 19 июня 1930 года.

на 192 листах.

Original interrogation list for Gnadenfeld 1930
Page 1

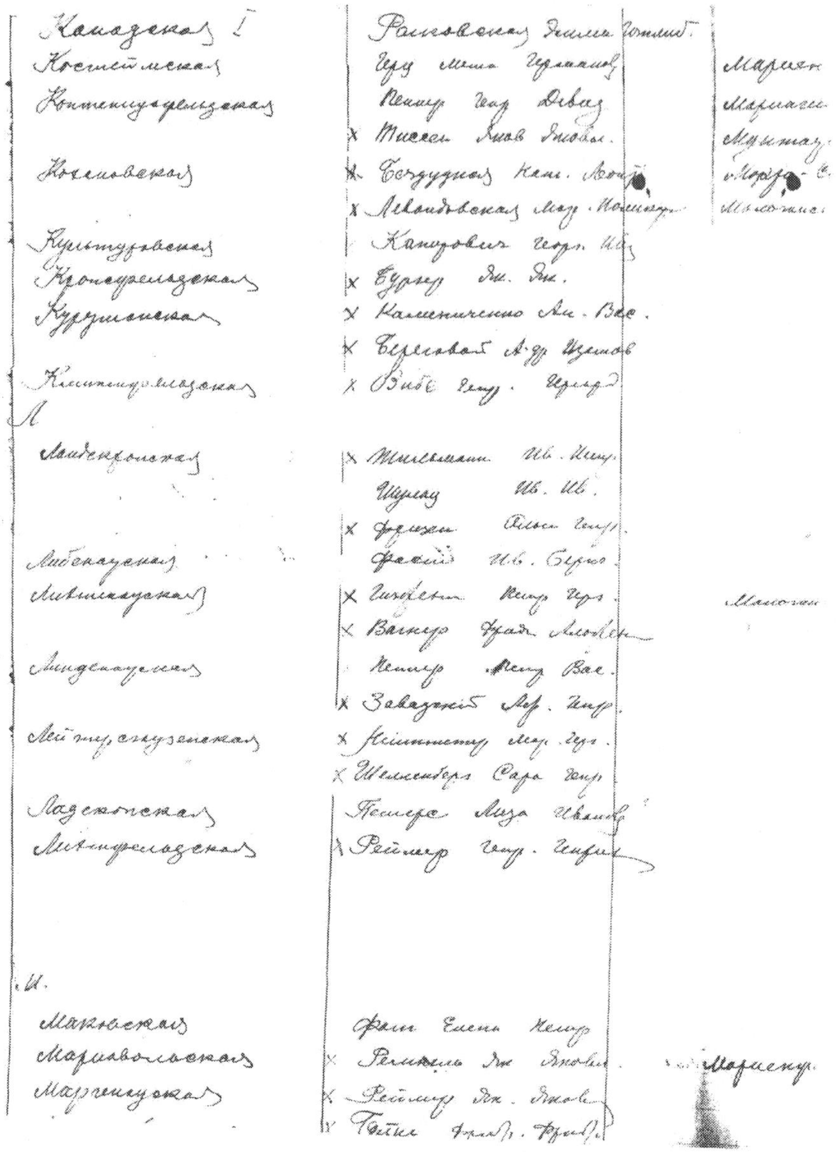

Original interrogation list for Gnadenfeld 1930
Page 2

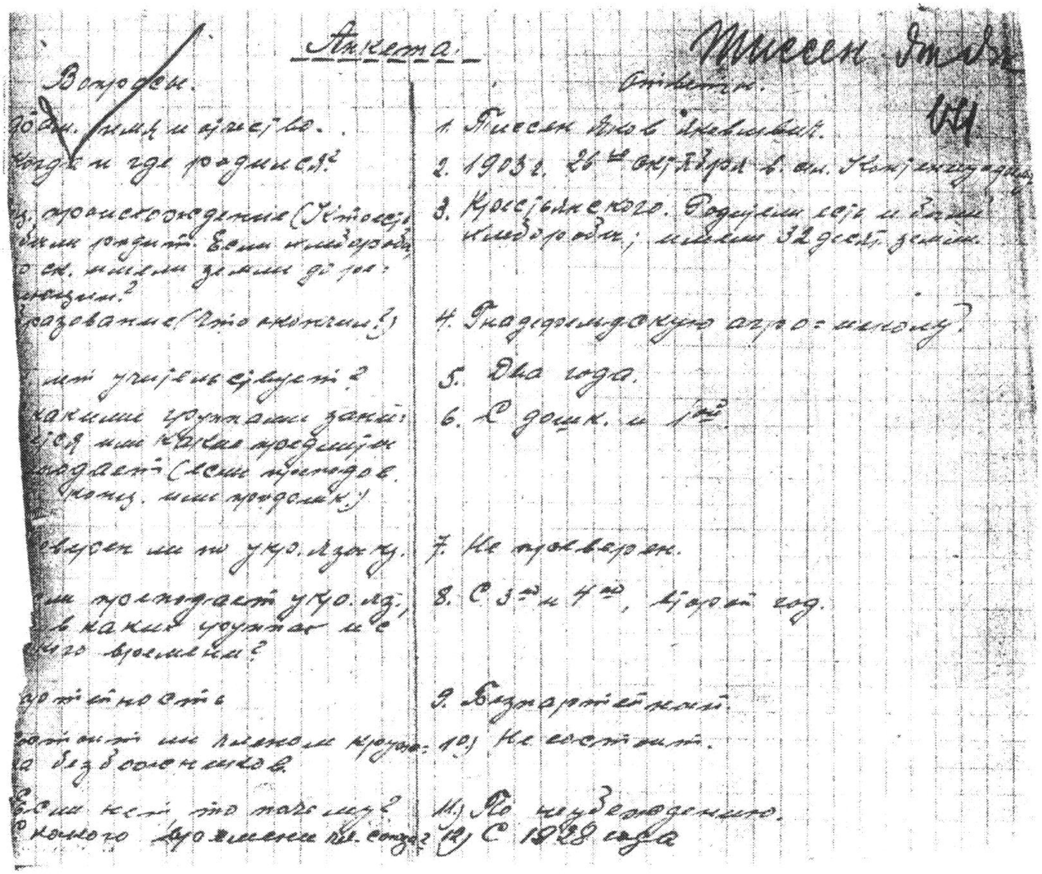

Original interrogation record for Teacher Jakob Thiessen 1930

Page 1

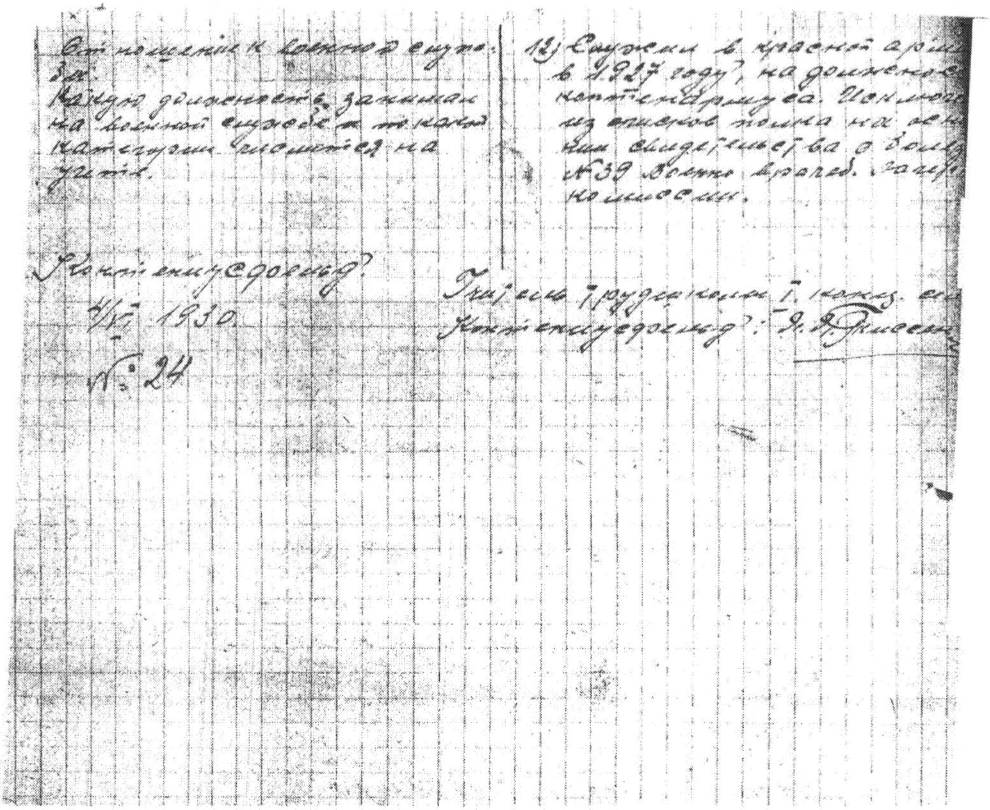

Original interrogation record for Teacher Jakob Thiessen 1930
Page 2

Translation of Front cover of Interrogation file

Orlov Village Labour Council, Peasant and Redauncy [?] Deputies Council, Orlov, Molochansk region, ???? District USSR

Protocol # 25

List of Questionnaires of the Studies of teachers in the Molochansk region

Begun 6 March 1930
Finished 19 June 1930
On 192 pages

[From the list of names in Alphabetical order according to the villages in the region]

Kontiniusfeld	
Penner	Heinrich
Tissen	Jacob Jakovich

Translation of Questionnaire form

Questions	Answers
1. Name and patronymic	1. Tissen Jacob Jakovich
2. When and where born	2. 1903, 26th October in Village Kontiniusfeld
3. Origin by birth, property before the revolution?	3. Peasant, Parents and 3 "Bread grains"They owned 32 desyatins of land.
4. Education (What did he finish)?	4. Gnadenfeld Agricultural School?
5. Teacher (Is he a teacher?)	5.Two years
6. With what sort of groups is he occupied, what subject matter?	6. With dosh to first [???]
7. Has he been tested for the Ukrainian Language?	7. Not tested.
8. Teaches in Ulkrainian, how long?	8. 3rd and 4th, for the second year.
9. Member of the Party?	9. Not a member.

10. Member of the circle of the godless?	10. Does not take his place
11. If not, then why not?	11. Is not convinced (lack of conviction)
12. Since when a member of SOP?	12. Since 1928
13. Attitude toward military	13. Served in the Red Army in 1927? In the capacity ...missing from the record of the regiment. Evidence... N.39 army doctor's camp commission!

Kontenusfeld,
Teacher in labour I Konz.
4 April 1930
Konteniusfeld J. J. Tissen

Questionnaire Penner Genrich Ivanovich 134

Name, patronymic: Penner Genrich [Heinrich] Ivanovich
Origin: Grain producer, 65 desyatin
Education; Gnadenfeld Zentral School, Berdyansk Real School
How many years active: 13 years. From 1917 - 1921 in the old gr. As a teacher in Rudnerweide ; from 1921 - 1930 in Gnadenfeld in 7 year
School [up to grade 7]
With what groups occupied and what subjects taught: Algebra (V, VI, VII) Physics (V, VI, VII), geometry (VI, VII) Russian lang. (VI, VII), ???? (V, VI, VII), geometry (IV)
Tested in Ukrainian: No
Teacher in what classes and for how long: taught in V and VI gr. from 1926 to 1929
Party membership: Not a member of the party
Member of the circle of the godless: no
If not, why not: up to the present, not decided
Since when a member of professional union: since June 1921
Attitude to military service: private, out of the military
Into what category does his attitude toward military service fall: [no answer]

No. 56

Signature: H. Penner

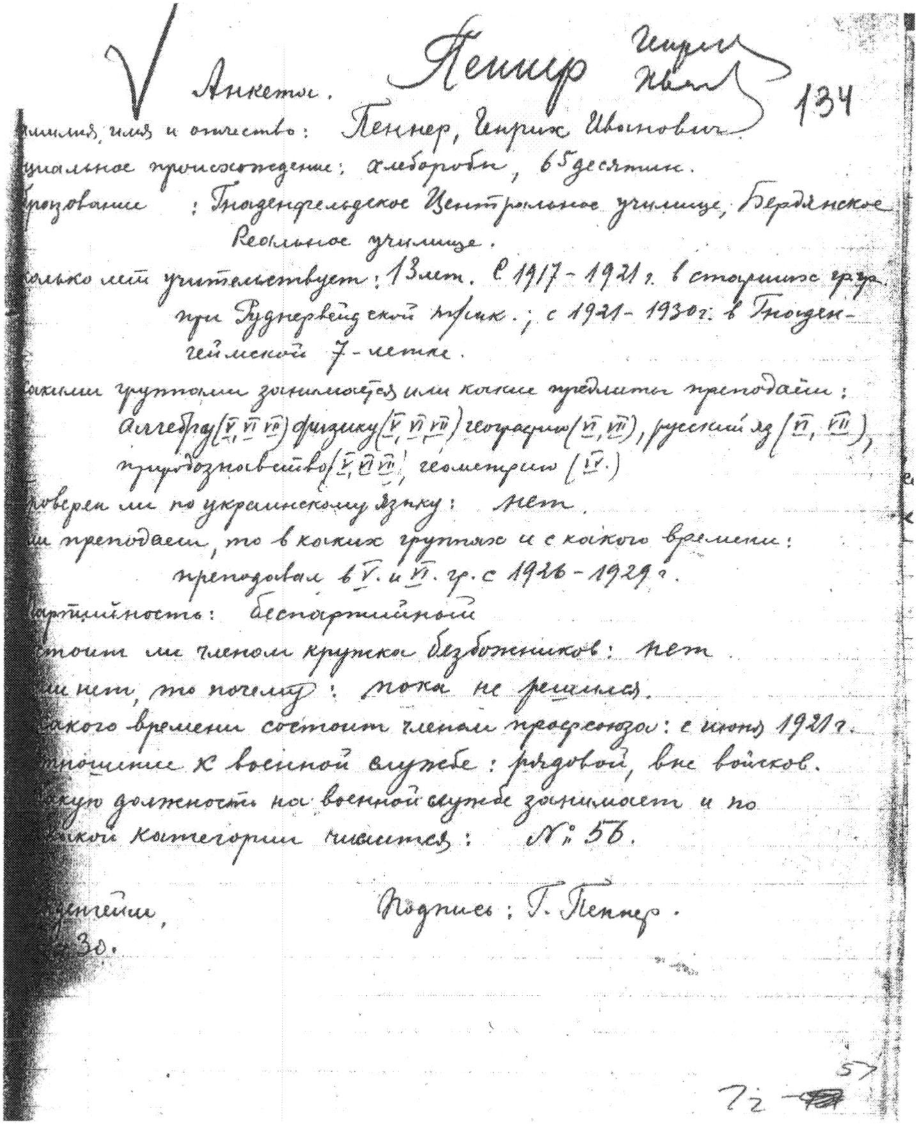
Анкета. Пеннер Генрих Иван. 134

амилия, имя и отчество: Пеннер, Генрих Иванович.

иальное происхождение: хлебороб, 65 десятин.

бразование: Гнаденфельдское Центральное училище, Бердянское Реальное училище.

олько лет учительствует: 13 лет. С 1917-1921 г. в старших гр. при Руднервейдской ... ; с 1921-1930 г. в Гнаденгеймской 7-летке.

акими группами занимается или какие предметы преподает: Алгебру (V, VI, VII) физику (V, VI, VII) географию (VI, VII), русский яз (VI, VII), природознавство (V, VI, VII), геометрию (IV.)

роверен ли по украинскому языку: нет.

и преподает, то в каких группах и с какого времени: преподавал в V. и VI. гр. с 1926-1929 г.

артийность: беспартийный

стоит ли членом кружка безбожников: нет.

и нет, то почему: пока не решился.

акого времени состоит членом профсоюза: с июня 1921 г.

тношение к военной службе: рядовой, вне войсков.

акую должность на военной службе занимает и по акой категории числится: №: 56.

денгейм, 30.

Подпись: Г. Пеннер.

57

72

Original interrogation record for Teacher Heinrich Penner 1930

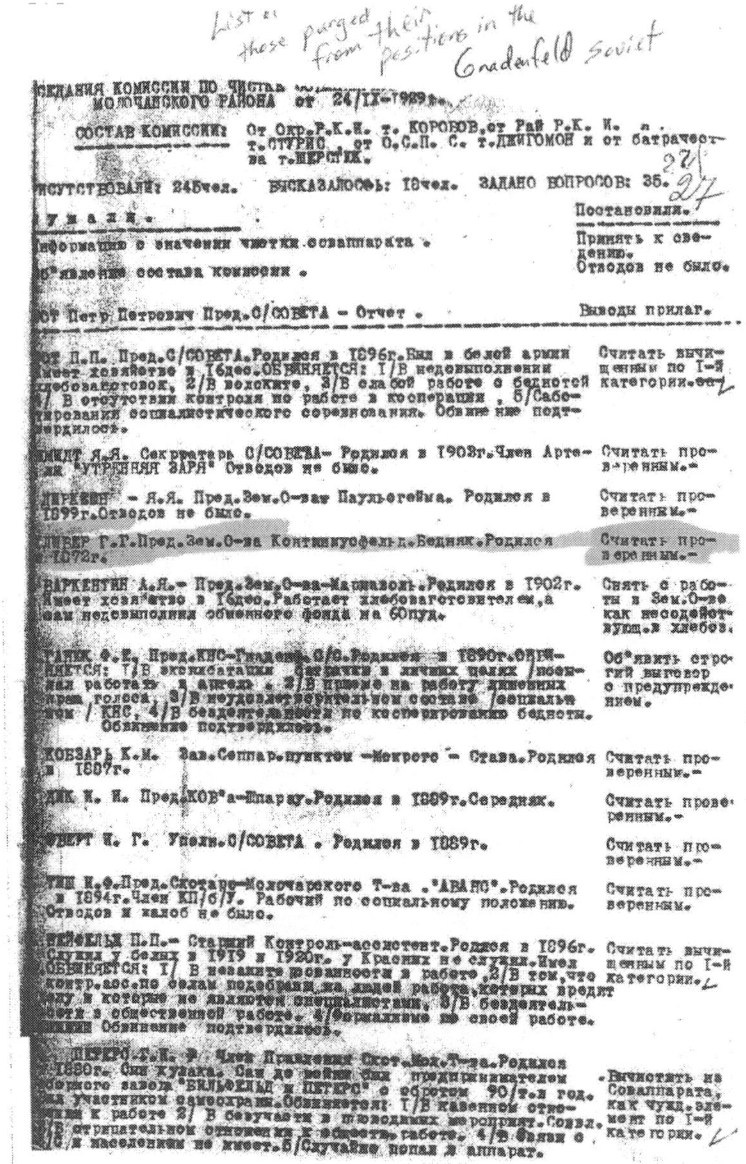

...СЕДАНИЯ КОМИССИИ ПО ЧИСТКЕ [illegible]
МОЛОЧАНСКОГО РАЙОНА от 24/IX-1929г.

СОСТАВ КОМИССИИ: От Окр.Р.К.И. т. КОРОБОВ, от Рай Р.К. И. т.СТУРИС, от О.С.П. С. т.ДЖИГОМОН и от батрачества т.ШЕРСТЮК.

ПРИСУТСТВОВАЛИ: 245чел. ВЫСКАЗАЛОСЬ: 18чел. ЗАДАНО ВОПРОСОВ: 35.

Слушали.	Постановили.
Информацию о значении чистки совапарата.	Принять к сведению.
Объявление состава комиссии.	Отводов не было.
...от Петр Петрович Пред.С/СОВЕТА – Отчет.	Выводы прилаг.
...от П.П. Пред.С/СОВЕТА. Родился в 1896г. Был в белой армии. Имеет хозяйство в 16дес. ОБВИНЯЕТСЯ: 1/В недовыполнении хлебозаготовок, 2/В волоките, 3/В слабой работе с беднотой 4/ В отсутствии контроля по работе в кооперации, 5/Саботировании социалистического соревнования. Обвинение подтвердилось.	Считать вычищенным по I-й категории.
...ШМИДТ Я.Я. Секретарь С/СОВЕТА – Родился в 1903г. Член Артели "УТРЕННЯЯ ЗАРЯ" Отводов не было.	Считать проверенным.
...ДИРКСЕН – Я.Я. Пред.Зем.О-ва Пауль-Гейма. Родился в 1899г. Отводов не было.	Считать проверенным.
...ЛИВЕР Г.Г. Пред.Зем.О-ва Континиусфельд. Бедняк. Родился в 1872г.	Считать проверенным.
ВАРКЕНТИН А.Я. – Пред.Зем.О-ва-Мариавоны. Родился в 1902г. Имеет хозяйство в 16дес. Работает хлебозаготовителем, а сам недовыполнил обменного фонда на 60пуд.	Снять с работы в Зем.О-ве как несодействующ. в хлебоз.
...ТАНИК Ф.Е. Пред.КНС-Гнаденф. С/С. Родился в 1890г. ОБВИНЯЕТСЯ: 1/В эксплоатации [illegible] в личных целях /посылал работать [illegible] 2/В приеме на работу лишенных права голоса, 3/В неудовлетворительном составе /социальном/ КНС, 4/В бездеятельности по кооперированию бедноты. Обвинение подтвердилось.	Объявить строгий выговор с предупреждением.
КОБЗАРЬ К.М. Зав.Сеппар.пунктом –Мокрого– Става. Родился в 1887г.	Считать проверенным.
...ДИК И. И. Пред.КОВ'а-Шпарау. Родился в 1889г. Середняк.	Считать проверенным.
...БЕРТ И. Г. Уполн.С/СОВЕТА. Родился в 1889г.	Считать проверенным.
...ТИШ И.Ф. Пред.Скотаро-Молочарского Т-ва. "АВАНС". Родился в 1894г. Член КП/б/У. Рабочий по социальному положению. Отводов и жалоб не было.	Считать проверенным.
...НЕЙФЕЛЬД П.П. – Старший Контроль-ассистент. Родился в 1896г. Служил у белых в 1919 и 1920г. у Красных не служил. Имел ... ОБВИНЯЕТСЯ: 1/ В незаинтересованности в работе, 2/В том, что контр.асс. по селам подобрал на людей работа, которых вредит [illegible] и которые не являются специалистами. 3/В бездеятельности в общественной работе. 4/Формализме по своей работе. Обвинение подтвердилось.	Считать вычищенным по I-й категории.
...ПЕТЕРС Г.Я. Член Правления Скот.Мол.Т-ва. Родился в 1880г. Сын кулака. Сам до войны был предпринимателем [illegible] завода "ВИЛЬФЕЛЬД и ПЕТЕРС" с оборотом 90/т.в год. Был участником самоохраны. Обвиняется: 1/В халатном отношении к работе 2/ В неучастии в проводимых мероприят. Соввл. 3/В отрицательном отношении к общественной работе. 4/В связи с [illegible] населением не имеет. 5/Случайно попал в аппарат.	Вычистить из Совапарата, как чужд. элемент по I-й категории.

List of local officials in Gnadenfeld area examined by the Soviet Commission 1929
(Page 1)

- 4 -

...жуналы.	Постановили.
...ЙДЕ К.К. Почталион. Родился в 1880г. Член Кулацкой ар-...али. Был. экспертник.	Снять с работы. 3? 30
...РИБ А.Ф. Зав. Мед. амбулаторией. Служащий. Родился в 1893г. ...ужил у белых.	Считать проверенн
...ИЛДСБЕРГ С. В. Аккушерка. Родилась в 1867г.	Считать проверенно
...НЦЕН И. Я. Зав. Сеппар. пунктом-Кантинауфельд. Родился в 1902г. Середняк.	Считать проверен.
...ТЕРС П.К. Зав. Сеп. пунктом-Паутель. Родился в 1868г. ...редняк.	Считать проверен.
...РЦ Ф.П. Зав. Сеп. пунктом Пауфельд Родился в 1874г. Середняк. Сотоит в Тракторном Т-ве.	Считать проверенным.-
...НЕЙФЕЛЬД П.Д. - Контроль-ассистент-Марновоь. Родился в 1901г. Был в Красной Армии. ~~Считать проверенным.~~	Считать проверен.
...ВЗЕ Г.Д. Зав. Сеп. пунктом Марновоь. Родился в 1877г. Середняк.	Считать проверен.
...НККЕР Г.Г. - Пред. С/С Александрсталь. Родился в 1904г. Был в Красной Армии. Обвиняется: 1/В бездеятельности, как ПредС/С и, как не отвечающий своему назначению. Обвинение подтвердилось.	Снять с работы, как несоответств. своему назначению.
... РЕГАН П. П-Контроль-ассистент. Шпардау. Родился в 1902г. В Армиях не служил. Работает в комиссии по содействию хлебозаготовки.	Считать проверенным.-
...ИБЕ Я.Я. - Зав. Сеппар. пунктом-Шпардау. Родился в 1905г. Состоит Пред. СОЗ"а. Отводов не было.	Считать проверенным.-
...ЭНС А.Я. Зав. Сепп. пунктом-Шпардау. Родился в 1891г. Работает Уполномоченным С/С.	Считать проверенным.-
...ДИРКСЕН И.К. - Контроль-ассистент-Шпардау. Родился в 902г. В армиях не служил. Отводов не было.	Считать проверенным.-
ПАЛЮХОВИЧ И.В. - Зав. Почтой - Гнаденфельд. В армиях не служил. РОдился в 1889г. Отводов не было.	Считать проверенным.
...РАПШ Я.Я. Почтольон. Родился в 1894г. В армиях не служил. В общественной работе участвует.	Считать проверенным.
...ШМИДТ И.П. - Почт. телегр. работник. В армиях не служил по религиозному убуждению. Родился в 1907г. Не член Союза.	Считать провер.
...РЕНСЛАБ В. Г. - Почтольон. Родился в 1903г. В армии не служил по религиозному убеждени. Имеет хоз-во в 8дес. земли.	Считать проверенным.-
...ЯНЦЕН Д.П. - Пред. ПКТ-Гнаденфельд. Родился в 11898г. ОБВИНЯЕТСЯ: 1/ В недовыполнении плана хлебозаготовок. 2/В неудовлетворительном подборе аппарата кооперации.	Считать проверенным, но объявить выговор за халатность.
...ЭЛИГЕР И.С. - Кассир П.К.Т. - Гнаденфельд. Родился в 1872г. ОБВИНЯЕТСЯ: 1/В грубости с пайщиками 2/Припиоке отпускаемых товаров, 3/Незаинтересованности в работе. Саботаж. 4/В подаче неправильных показаний в анкете/написал "Батрак", а фактически был "Управляющим"/.	Вычистить по 1-категории, как чуждый элемент.
...НЕЙФЕЛЬД Г.Я. - Закупщик хлеба. Родился в 1876г. Был исключен из Профсоюза. Специальных знаний, как хлебозаготовитель не имеет.	[illegible] Снять с работы как не специа- -по хлебному ...у.-
...ФОТ Г.Я. - Приказчик П.К. т. Родился в 1911г. Член Союза с 1927г. ОБВИНЯЕТСЯ: В грубом обращении с русскими женщинами 2/В слабой заинтересованности в кооперативном деле. 3/В игнорировании общественной работы.	Снять с работы в кооперации за грубость и бездеятельность
...НИЦЕН П.П. Зав. Маг. П. К. т. Родился у 1886г. В армиях не служил. Имел торговлю в 1913г. Работой не интересуется ...ти в работы. 2/С массой не связан, исключая кулак...	Вычистить по 1-й категории, как чуждый элемент.

8h-12

List of local officials in Gnadenfeld area examined by the Soviet Commission 1929

(Page 2)

Postcard 1930 from Kontiniusfeld

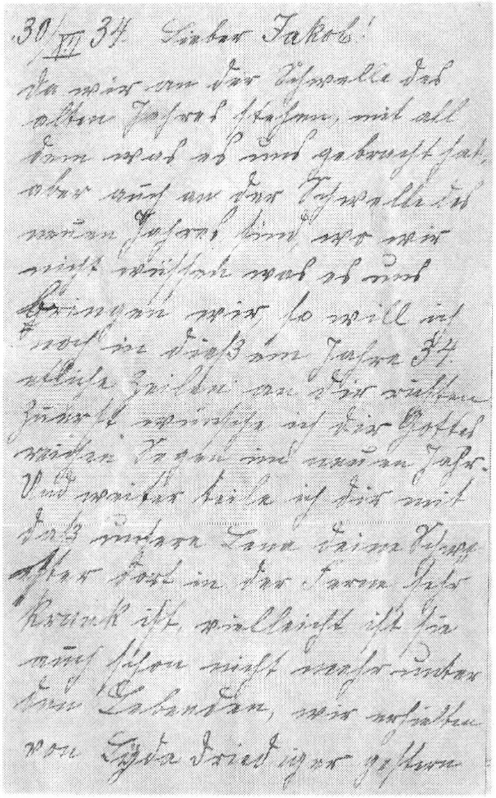

30/XII 34. Lieber Jakob!
Da wir an der Schwelle des
alten Jahres stehen, mit all
dem was es uns gebracht hat,
aber auch an der Schwelle des
neuen Jahres sind wo wir
nicht wissen was es uns
bringen wird, so will ich
noch in diesem Jahre 34
etliche Zeilen an Dir richten.
Zuerst wünsche ich Dir Gottes
reichen Segen im neuen Jahr.
Und weiter teile ich Dir mit
daß unsere Lena [illegible]
[illegible] Dorf in der Farm sehr
krank ist, vielleicht ist sie
auch schon nicht mehr unter
den Lebenden, wir erhielten
von [illegible] Friesen gestern

An original letter and envelope from Kontiniusfeld

Made in United States
Orlando, FL
04 November 2026

90200714R00166